I0055585

Quantum
Bio-Informatics VI

From Quantum Information to Bio-Informatics

Proceedings of Quantum Bio-Informatics 2014

QP–PQ: Quantum Probability and White Noise Analysis*

Managing Editor: W. Freudenberg
Advisory Board Members: L. Accardi, T. Hida, R. Hudson and
K. R. Parthasarathy

QP–PQ: Quantum Probability and White Noise Analysis

*For the complete list of the published titles in this series, please visit:
www.worldscientific.com/series/qp-pq

QP–PQ
Quantum Probability and White Noise Analysis
Volume XXXI

Q*uantum*
B*io-*I*nformatics* VI

From Quantum Information to Bio-Informatics

Proceedings of Quantum Bio-Informatics 2014

Tokyo University of Science, Japan   *23 – 25 October 2014*

Editors

Luigi Accardi

Università di Roma "Tor Vergata", Italy

Wolfgang Freudenberg

Brandenburgische Technische Universität Cottbus, Germany

Noboru Watanabe

Tokyo University of Science, Japan

World Scientific

NEW JERSEY · LONDON · SINGAPORE · BEIJING · SHANGHAI · HONG KONG · TAIPEI · CHENNAI · TOKYO

Published by

World Scientific Publishing Co. Pte. Ltd.

5 Toh Tuck Link, Singapore 596224

USA office: 27 Warren Street, Suite 401-402, Hackensack, NJ 07601

UK office: 57 Shelton Street, Covent Garden, London WC2H 9HE

British Library Cataloguing-in-Publication Data
A catalogue record for this book is available from the British Library.

QP-PQ: Quantum Probability and White Noise Analysis — Vol. 31
QUANTUM BIO-INFORMATICS VI
From Quantum Information to Bio-Informatics
Proceedings of Quantum Bio-Informatics 2014

Copyright © 2020 by World Scientific Publishing Co. Pte. Ltd.

All rights reserved. This book, or parts thereof, may not be reproduced in any form or by any means, electronic or mechanical, including photocopying, recording or any information storage and retrieval system now known or to be invented, without written permission from the publisher.

For photocopying of material in this volume, please pay a copying fee through the Copyright Clearance Center, Inc., 222 Rosewood Drive, Danvers, MA 01923, USA. In this case permission to photocopy is not required from the publisher.

ISBN 978-981-121-782-1 (hardcover)
ISBN 978-981-121-783-8 (ebook for institutions)
ISBN 978-981-121-784-5 (ebook for individuals)

For any available supplementary material, please visit
https://www.worldscienti ic.com/worldscibooks/10.1142/11760#t=suppl

PREFACE

This volume collects papers that were discussed during the international workshop of quantum bio-informatics held at the QBIC of Tokyo University of Science. This is a special issue dedicated to Prof Masanori Ohya, founder of QBIC and initiator of this series of conferences.

The purpose of the workshop is to contribute to the creation of a deeper interaction among different scientific disciplines like mathematics, physics, information and life sciences. In particular we look for a new paradigm for the role of quantum theory in the connection between information science and life science. We had a lot of fruitful discussion with several active researchers in various fields such as mathematics, physics, information and biology coming from all over the world. In this workshop, particular attention is given to quantum entanglement, simulation of bio-systems, brain function, quantum like dynamics and adaptive systems. Most of the speakers gave care to the relation between their own topics and the mystery of life.

All papers in this volume have been refereed. Their contents are related to one of the following subjects:

(1) Mathematics of Cryptography and its related topics in homomorphic encryption system
(2) Quantum algorithm based on adaptive dynamics
(3) Quadratic Boson fields and quantum quadratic stochastic operators
(4) Quantum entropy, compound states and quantum channels
(5) Information dynamics and entropic chaos degree
(6) Quantum states on Fock space
(7) Open systems and quantum dynamical entropy
(8) Stochastic dynamics and white noise analysis
(9) Local gauge invariance and symmetry breaking
(10) Situation in quantum foundations after quantum probability
(11) Protein structure and Brownian dynamics in silico biology
(12) Alignment of sequences and protein coding genes

Luigi Accardi
Wolfgang Freudenberg
Noboru Watanabe

MASANORI OHYA:
A PIONEER OF QUANTUM INFORMATION

LUIGI ACCARDI

Centro Interdipartimentale Vito Volterra
Università degli Studi di Roma "Tor Vergata"
Facoltà di Economia
Via Columbia, 2, Italy
E-mail: accardi@volterra.mat.uniroma2.it – http://volterra.mat.uniroma2.it

1. Introduction

Masanori Ohya begun his scientific carrier under the guidance of Isaharu Umegaki who was one of the founders of the famous Japanese school of operator algebras. Among other important contributions to this field (probably the best known of which is the operator–theoretic notion of *conditional expectation*), Umegaki first extended the notion of Kulback-Leibler relative information defining the relative entropy in the general framework of semi–finite von Neumann algebras. It is interesting to read the short preface written by Roman Ingarden to the Selected Papers of Umegaki ([34], p. IX), to realize how the scientific programme underlying this development, i.e. that of a mathematical approach to the quantum aspects of information theory combined with its physical aspects, was immediately perceived as interesting and fruitful by a group of mathematician and physicists that in those times was small, but as we now know fated to conspicuously increase. It is not by chance that some of the early papers of Ohya were on quantum relative entropy and were written in collaboration with Fuimio Hiai. This subject would accompany both scientists throughout their scientific carrier even if in slightly different directions: Hiai more concentrated on pure mathematics, Ohya with a constant attention to applications in different fields. Starting from those early years, the core inspiration for M. Ohya has undoubtedly been quantum information, of which he can be considered with Roman Ingarden and Andrei Kossakowski as one of the founding fathers and pioneering figures who anticipated of several years the impetuous development of quantum information that has taken place in the past 20 years. Their book [3] is not only a precious source of information for anybody interested in quantum information and its connections with equilibrium and non–equilibrium quantum statistical mechanics, but it also marks a milestone in the long–time scientific collaboration between the Torun and the Noda groups in quantum information and open systems, a collaboration that actively continues also nowadays after the demise of both its initiators.

Classical information theory made its transition from a purely engineering and technological level to a well established mathematical theory, based on non trivial theorems, in the 1960's with the work of C. Shannon who, developing earlier intuitions of N. Wiener, introduced the notion of entropy as a measure of information. Shannon's work inspired Khintchine's pioneering work on the mathematics of information theory and, most of all, Kologorov's work on entropy and information (in particular Kolmogorov's notion of dynamical entropy, see further discussion).

The idea of a quantum extension of Shannon's results naturally arose in the physical literature in the context of quantum signal processing and, already in the late 1960's, was discussed by several authors (among whom: Gordon, Lax, Louisell, ...) and nowadays quantum information, enriched by the engineering appeal of the quantum technology programme, has become a widely practiced scientific discipline and the object of huge investments from all industrialized countries.

But in the early 1980's these investigations were cultivated by a handful of pioneer researchers and the influence of M. Ohya's former advisor, Professor Umegaki, surely played a decisive influence in his scientific development. In fact the barycentre of Masanori Ohya's scientific interest can be described as follows: *use of advanced mathematical techniques for the development of classical and quantum information theory and its applications to physics, life sciences economy, sociology and even psychology.*
The effort to develop the quantum generalization of a line of research naturally leads to a new and deeper comprehension of its classical aspects and this stimulates in many scientists an effort to apply the new techniques and ideas to different branches of science.
This happened with M. Ohya and naturally led him to become actively involved in applications to life science, where the mechanism of elaboration and transmission of information plays a crucial role.
The constant attention to the concrete application to experimental data of the theoretical tools developed is at the basis of the intensive collaboration

between Ohya's group and different teams working in bio–medical sciences. Some of the notions introduced by Ohya in information theory, in particular quantum compound state, mutual entropy, entropy evolution rate and entropic chaos degree, have found applications not only in mathematics, but also in the study the dynamical evolution of HIV virus in terminal patients (joint work with S. Miyazaki and K. Sato [23]).

The first Ohya's paper dedicated to biology was devoted to the mathematical and statistical study of alignment of DNA sequences in human genome and protein sequences [27]. This paper was the starting point of a brilliant research line that was pursued by Ohya's former student Satoru Miyazaki, who became a world renowned expert in this field, has created his own group in Tokyo University of Science for computer analysis of DNA sequences which is bringing valuable contributions to this important research topic and is now Dean of the Faculty of Pharmaceutical Sciences of the Graduate School of Pharmaceutical Sciences in this university. Later, in a joint paper with T. Hara and K. Sato [7], Ohya further contributed to this line of research finding a method for sequence alignment which seems to produce the most accurate predictions among similar methods.

This collaboration with the bio–medical area has led among other results, to a joint patent with the medical school of Tokyo University of Science concerning drug particle movements in blood (a description of this model is contained in [5], p. 690).

Chronologically, the latest development of this intellectual adventure was the merging of these two lines of research into the quantum bio–information program, a bold attempt to investigate the role of quantum physics and quantum information theory, in life sciences.

Since quantum mechanics represents the deepest level of development of contemporary physical sciences and since it has been now recognized that life is the most sophisticated form of physical elaboration, preservation and transmission of information, it is natural to conjecture that such role should exists and be relevant.

However even with the tremendous advance of experimental bio–physics in the past 30 years, the order of magnitude of microscopic quantum phenomena is still far from experimental reach of bio–physicists and it is natural to look for confirmation of this natural conjecture in the more restricted domain of *macroscopic quantum phenomena*. It is exactly from this domain that now begin to emerge the first confirmations of this conjecture.

One of the most interesting such confirmations comes from the increasing evidence of the role of quantum effects in one of the most important biological phenomena, namely Photosynthesis where the order of magnitude of transport of excitons from sensors to the reaction center has not found a convincing classical explanation and seems to require the presence of quantum effects [33]. Several models have been proposed by different groups in the world to explain this fact and probably the most convincing one is the Aref'eva–Kozyrev–Volovich model [31] which starts from first physical principles and applies the stochastic limit technique deducing from it a master equation which accounts for the above mentioned transport phenomenon. Also in this direction Ohya was a precursor and the series of QBIC (Quantum Bio–Informatics and Computation) created a communication channel among physicists, mathematicians, information theorists, and biologists was at the origin of this line of research which continues to involve Ohya's group with the collaboration between Satoshi Iriyama and Volovich's group and recently has been enriched by the contributions of Roberto Quezada's mexican group and Skander Hachicha Tunisian group.

Any deep scientific advancement is based on a mix of established knowledge and speculation and the role of pioneers is precisely to move the first steps in the direction of transforming speculations and hopes into realizations and facts.

One of the main merits of M. Ohya consists in having understood that such an ambitious scientific program could not be handled by a single individuum: the success of the program required coordination of the energies and enthusiasm of young generations with a solid network of mature experts in different fields and different countries.
He succeeded remarkably well in both directions, first of all with an intensive activity as educator which produced hundreds (literally) of students some of whom now have achieved preeminent positions in different branches of science, of industry and of administration; second, with a careful selection of collaborators from all over the world, which has made Tokyo University of Science one of the main international centers of quantum information theory since almost three decades.

One of the highlights of the national and international network built by M. Ohya is the success of the international journal *Open Systems and Information Dynamics*, founded by Ohya and Ingarden and published for several years by the Torun University group before passing to World Scientific, which in a few years has achieved an impact factor higher than many much older scientific journals, and whose foundation was based on an early intuition of M. Ohya anticipating of several years the need for a quantum information journal. In the past few years many such journals have been founded, dealing with various aspects of quantum information, however this recent proliferation has not altered the position of qualitative excellence of OSID in this field.

The term *Information Dynamics* was coined by M. Ohya who, in his selected paper ([9], p. 9) describes it as ' ... *a synthesis of the dynamics of state change and the complexity of states* ... . He realized that the notion of *complexity of a quantum state* can be quantified, as in the classical case, in several different ways and extended to the quantum case several of the *complexity indexes* introduced by Kolmogorov which can be considered as variations (topological, capacity, dynamical, ...) of the notion of entropy. In addition he introduced new measures of complexities such as transmitted complexity, entropic chaos degree, and applied them to a multiplicity of phenomena in several different disciplines such as optical communication processes, genetic matrix for genome process, optical illusion, time series in different economics data, ...

While the creation of a journal devoted to a new topic requires vision and management capacities, the writing of a monograph such as *Quantum Entropy and his use*, by M. Ohya and D. Petz, requires a level of scholarship that is rarely found in thousands of scientific books appearing nowadays. It is not by chance that this Springer book, now at its second edition, is well established since decades as one of the classics of quantum information theory.

In addition to this the series of annual workshops on *Quantum Information Theory and Open Systems*, organized by M. Ohya in collaboration with Izumi Ojima since 1992 at the Kyoto Research Institute of Mathematics (RIMS), were the first periodical meeting point for the Japanese researchers interested in the mathematical aspects of quantum information.

These achievements parallel an impressive scientific production, marked by some highly original discoveries, as well as the active participation in the editorial board of several international journals, an intense publishing activity including several books in different areas, and in particular the publication of the highly successful *Encyclopedia of Information Sciences*.

Each of these activities would be more than enough for a normal academic carrier. Therefore it is truly remarkable how M. Ohya could combine all of them and make them compatible with the additional duties of Dean of the Graduate School of Science and Technology, Director of the Education Research Center for Information Science and Technology, Member of the Scientific Committee of the International Institute for Advanced Studies, trustee of Tokyo University of Science, which he served during several years, sometimes in mutual coincidence.

Nowadays the pressure of specialization is very high. One of its psychological effects is that many scientists remain captured in a narrow horizon and become existentially unable to appreciate results beyond such a narrow horizon. There is a sociological counterpart of this effect which makes some people unable to communicate, or to enjoy communication, with people outside a narrowly outlined community. Many scientists, even of very high technical level, do not succeed in escaping such a trap.
Therefore those, like M. Ohya, who succeed in maintaining a broad vision of science while struggling to realize his own visions and scientific programmes, without abdicating the luxury of having a wide human and scientific taste which allow him to sincerely acknowledge scientific results even if outside of his sphere of direct engagement or of academic alliances, provide illuminating examples for the future generations, so much precious as they are rare.

2. Some more specific comments on some of M. Ohya's scientific results

One of the earliest, and still among the most important, of M. Ohya's contributions to quantum information theory, is his quantum extension of the notions of mutual entropy and compound channel [29], whose classical counterparts play a basic role in Shannon's approach to information theory.
This extension belongs to the class of probabilistic results whose generalization from classical to quantum cannot be reduced to a mere translation because the specific quantum features, namely the restricted existence of joint probabilities, make such a translation impossible: new ideas were needed and the fact that, after almost 30 years no plausible alternative candidate for this notion has emerged suggests that Ohya's notion of mutual entropy will stably remain in the future as the appropriate quantum extension of this basic pillar of classical information theory.

A few years later the two notions of compound channel and of transition expectation were combined into the unifying notion of *lifting* [26] which has now found several applications in different fields of mathematics, physics and quantum information.

Inspired by Shannon's work, Kolmogorov introduced entropic type quantities as a measure of complexity and as a tool in approximation theory. These were generalized by M. Ohya to the quantum case [21] and also in the classical domain with the introduction of the notion of *chaos degree*. This notion found unexpected experimental support in econometrics, (joint papers of M.O. and T. Matsuoka) and in medicine (joint papers of M.O. and K. Saito).

Kolmogorov also introduced a notion of dynamical entropy and proved that it provided the first example of invariants for dynamical systems finer than the spectral invariants introduced by Halmos and von Neumann.
The fundamental step in Kolmogorov's construction was to associate a family of finite–state Markov chains to the given reversible dynamical system, defined as a measure–preserving 1–parameter group of measure preserving, bi–jective transformations of a measure space. The calculation of the entropy of these chains can be done explicitly and taking the supremum over all these chains, he obtained a number depending only on the dynamical system. The surprising fact is that, as shown by Kolmogorov, this supremum in some cases can be explicitly calculated.

Several quantum generalizations of Kolmogorov dynamical entropy have been proposed, but these constructions were very complicated and most of all **they were lacking this direct and canonical connection with the theory of Markov chains**.
This motivated the paper [25] where, starting from a quantum dynamical system and extended to this case Kolmogorov's original construction using quantum Markov chains rather than classical to extend step by step the original Kolmogorov construction to the quantum case.
The computation of the entropy of a quantum Markov chain is not so simple as in the classical case, however in several concrete examples it was possible to calculate this entropy or at least to produce lower and upper bounds for it [24]. The construction was later extended to irreversible dynamical systems by Ohya, Kossakowski and Watanabe [18].

Another anticipating intuition of M. Ohya was the recognition that the theoretical studies on quantum information and communication could find concrete realizations and applications in quantum computer and quantum cryptography disciplines which in the late 1990's begun to gather momentum in scientific environments.

Thus he himself and his group begun to actively work in this field long before it became a fashionable scientific trend.

The **channel theoretical formulation of teleportation**, that M. Ohya had developed in collaboration with Kei Inoue, Hiroki Suyari and Noboru Watanabe [20], [22], can be considered as the first important contribution of Ohya's group to quantum computer. This formulation has now become the standard mathematical formulation of teleportation, universally adopted in the literature even if the public acknowledgement to its initiators is by far not as frequent as its use.

While the original formulation of this notion, due to Bennet and Brassard, was limited to a very specific example and did not suggest any insight into its general structure, Ohya's channel–theoretical formulation really captured the mathematical essence of teleportation thus opening the way to its realization in arbitrary dimensions.

On this problem several groups were independently working in different parts of the world but, at that moment, it was still open. The first constructive solution of this problem was given in the paper [15] which also included the first proof of the fact that the teleportation scheme **must** be based on a maximally entangled state in a given basis. Such a result was quite unexpected at that time and was later rediscovered and generalized in various ways by several authors among whom R. Werner.

The extension of this result to infinite dimension, where maximally entangled pure states do not exist, was a nontrivial mathematical challenge motivated by the experiments physicists begun to do in (which were based on coherent states). Such challenge eventually found its solution with the introduction of the notion of *incomplete teleportation* in the papers [16], [17]. The restriction to coherent states and their superpositions was finally overcome in the paper [10] thus allowing in particular the unification of the finite and infinite dimensional case.

Another outstanding result, obtained by Ohya in collaboration with Masuda [19], is the possibility to solve the SAT problem in polynomial time using quantum computation techniques. In classical complexity theory it is known that the fact that the SAT problem is polynomial implies the coincidence of the P and NP classes of problems. Thus Ohya's result proves that, if quantum computation is allowed to enter the game, this long standing problem of complexity theory has a positive solution, at least in the probabilistic sense (which is a common feature of all quantum algorithms). The proof is based on the same assumptions used in the proof of the famous Shor factorization algorithm, i.e. on one side the possibility of constructing a unitary operator, which from the point of view of physics means the Schrödinger evolution of a quantum system, implementing a given function (in a sense now standard in quantum information), on the other side on the so called Lüders–Zumino formula in quantum measurement theory. A deeper discussion of the SAT result should be framed within a general analysis of these two assumptions. This will be done in a future paper.

Therefore, from the point of view of mathematics and physics, the status of the two above mentioned results is absolutely the same.

The complete realization of this algorithm had still to overcome a problem: one could not exclude the possibility that a positive solution (i.e. one guaranteeing satisfiability) would appear with such a small probability to be indistinguishable from a negative one. The proposal to amplify these probabilities by coupling the output of the quantum computation with a classically chaotic system, was advanced by Ohya and Volovich. The amplification of microscopic effects, by using them as triggers of chain reactions leading to macroscopically observable phenomena, is at the basis of a large class of measurements involving atomic and sub–atomic particles. The new element here was the appeal to chaotic dynamics.

A different, purely quantum method was later proposed in [12] combining the new technique of stochastic limit of quantum systems (see [14]) with the new idea of *state–adaptive dynamics* and applies both of them to the quantum state obtained as output of the Ohya–Masuda algorithm.

This combination is used to construct a state dependent interaction (hence the name *state–adaptive dynamics*) capable of driving a system to a stationary state. Thus, by discriminating the limit stationary states, one can discriminate between a positive or negative solution of the SAT problem.

The idea introduced in the solution of this specific problem had an influence in the last part of the scientific activity of M. Ohya because it unveiled a new aspect of *adaptive dynamics*. The notion of adaptive dynamics was used in control theory (see [32]) in the sense that the control vector corrects the trajectory of a dynamical using as input the feed–back that shows the discrepancy between the actual trajectory and the target, i.e. the desired position at the end of the trajectory. Studies in the foundation of quantum mechanics had led to a completely different notion of adaptedness in which the local interactions introduce a modification of the trajectory of a dynamical system not by an external control parameter, but by a change of the law of motion induced by a new term in the interaction Hamiltonian that was not present in the original Hamiltonian. This new type of dynamics was used to construct a classical, deterministic, reversible dynamical system that violates the Bell inequalities [30]. In the case of the SAT problem, the output is a pure state of a quantum system, so the idea came naturally to consider an interaction Hamiltonian that depends not on an observable, but on a state of a system. Observable–dependent Hamiltonian interactions are common in physics (think of the spin–boson Hamiltonian) but to my knowledge state–dependent Hamiltonian interactions appear only in non–linear modifications of the Schrödinger or Heisenberg equations of motion. On the other hand the idea that an interaction could depend of the state of a system is very natural from the point of view of physics. From the mathematical point of view, as already emphasized in von Neumann's monograph [35], there is no difference because any pure state can be considered as an eigen–vector of an Hermitean operator, i.e. an observable in the physical interpretation. However from the point of view of concrete physical implementation the situation is different because some pure states appear clearly as eigen–vectors of well defined, experimentally measurable observables, some do not (think e.g. of a coherent state). Therefore the problem of physical implementation of a state dependent interaction cannot be solved on a purely mathematical ground and should be left to the cleverness of experimentalists. One can only insist on the fact that such interactions are physically natural and therefore there should be a way to experimentally implement them.

M. Ohya perceived the novelty of this new class of dynamical systems and begun a systematic development of a general theory of *adaptive dynamical systems* that includes both the *observable–adaptive* (chameleon effect) and the *state–adaptive* systems introduced in the stochastic limit approach to the SAT problem (see the paper [6] at the beginning of his Selected Papers where this program is clearly described).

The new notion of *entropic chaos degree*, introduced by M. Ohya and T. Matsuoka [13] provides a new criterium for entanglement, much easier to verify in concrete cases than previously introduced and more widely quoted criteria. This was applied in [11] to prove the entangled nature of certain quantum Markov chains which could not be proved with previous methods.

The book [5] represents a synthesis not only of the author's researches in the field of quantum computation, but also of those of a multiplicity of other groups as well as a richly documented and smoothly readable introduction to the topics mentioned in the title. There are few doubts on the fact that this book will constitute a stable reference and source of inspirations for those interested in the path leading from advanced and abstract mathematics to life sciences passing through the multi–faced aspects of classical and quantum information theory.

References

Books

1. Selected papers of M. Ohya, World Scientific (2008)
2. Ohya M., Petz D.: Quantum entropy and its use, Springer, Texts and Monographs in Physics (1993)
3. R.S. Ingarden, A. Kossakowski, M. Ohya: Information Dynamics and Open Systems. Classical and quantum approach, Kluwer (1997)
4. M. Ohya: Mathematical Foundation of Quantum Computer, Maruzen Publ. Company (1998)
5. M.Ohya and I.Volovich: Mathematical foundation of quantum information and computation and its applications to Nano– and Bio–systems, TMP-series, Springer-Verlag (2011)

Papers

6. M.Ohya, Adaptive dynemics and it s applications to chaos and NPC problems, in: Selected papers of M. Ohya, World Scientific (2008) 1–36
7. T.Hara, K.Sato, M.Ohya, MTRAP: Pairwise sequence alignment algorithm by a new measure based on transition probability between two consecutive pairs of residues, BMC Bioinformatics 2010,
8. L. Accardi, A. Khrennikov, M. Ohya, Quantum Markov Model for Data from Shafir-Tversky Experiments in Cognitive Psychology, Open Systems and Information Dynamics, 16 (4) (2009) 441-443
9. M.Ohya Adaptive Dynamics and Its Applications to Chaos and NPC Problem, QP-PQ:Quantum Prob. White Noise Anal. 21, Quantum Bio-Informatics, World Sci. Publishing, (2008) 181-216
10. A.Kossakowski, M.Ohya New Scheme of Quantum Teleportation, Infinite Dimensional Analysis, Quantum Probability and Related Topics, 10 (3) (2007) 411-420

11. L. Accardi, T. Matsuoka, M. Ohya: Entangled Markov chains are indeed entangled, Infinite Dimensional Analysis, Quantum Probability and Related Topics 9 (2006) 379-390

12. Luigi Accardi, Masanori Ohya: A stochastic limit approach to the SAT problem, Open systems and Information Dynamics, 11 (3) (2004) 219-233

13. M. Ohya, T. Matsuoka: Quantum Entangled State and Its Characterization, Foundation and Probability and Physics-3, AIP Proceedings 750 (2005) 298-306

14. Accardi L., Lu Y.G., Volovich I.: Quantum Theory and its Stochastic Limit, Springer Verlag (2002)

15. Luigi Accardi, Masanori Ohya: Teleportation of General Quantum States, Quantum Information, T. Hida, K.Saito (eds), World Scientific (1998), 57-70; Voltera Center preprint, (1998).

16. K-H.Fichtner and M.Ohya (2001) Quantum teleportation and beam splitting, Commun. Math. Phys., 225, 67-89.

17. K-H.Fichtner and M.Ohya (2001) Quantum teleportation with entangled states given by beam splittings, Commun. Math. Phys., 222, 229-247.

18. Ohya M., Kossakowski A., Watanabe N.: Quantum dynamical entropy for completely positive map, Infinite dimensional analysis, quantum probability and related topics, 1 (2) (1999) 267–282 Preprint (1998)

19. Ohya M., Masuda N.: NP problem in quantum algorithm, Open Systems and Information Dynamics, Vol.7, No.1, 33-39, 2000. arXiv:quant–ph/9809074 v2 13 dec (1998)

20. Ohya M., Watanabe N.: On the mathematical treatment of the Fredkin–Toffoli–Milburn gate, Physica D 120 (1998) 206–213

21. Ohya M.: Complexities and their applications to characterization of chaos, Int. Journ. of Theort. Phy., 37 (1998) 495

22. Inoue K., Ohya M., Suyari H.: Characterization of quantum teportation processes by nonlinear quantum channel and quantum mutual entropy, Physica D, 120 (1998) 117-124 SUT Preprint (1997)

23. Sato K., Miyazaki S., Ohya M.: Analysis of HIV by entropy evolution rate, Amino Acids, 14 (1998) 343–352 Invited talk to the: International Conference on quantum information and computer, Meijo University 1998

24. Accardi L., Ohya M., Watanabe N.: Dynamical entropy through quantum Markov chains, Open Systems and Information Dynamics 4 (1997) 71–87

25. Accardi L., Ohya M., Watanabe N.: Note on quantum dynamical entropies, Reports on mathematical physics, 38 N. 3 (1996) 457–469

26. Accardi L., Ohya M.: Compound channels, transition expectations and liftings, Applied Mathematics Optimization 39 (1999) 33–59 Volterra preprint N. 75 (1991)

27. M. Ohya: A mathematical analysis of DNA sequences, Symp. Appl. Funct. Anal. 8 (1985) 36–47

28. M.Ohya: Entropy Transmission in C*-dynamical systems, J. Math. Anal.Appl., 100 (1) (1984) 222-235

29. Ohya M.: On compound state and mutual information in quantum information theory, IEEE Trans. Information Theory, 29 (1983) 770–777

Other cited references

30. Accardi L., Imafuku K., Regoli M.: On the EPR-Chameleon Experiment Infin. Dimens. Anal. Quantum Probab. Relat. Top. (IDA-QP) 5 (1) (2002) 1-20 quant-ph/0112067
31. I. Ya. Aref'eva, S. V. Kozyrev, I. V. Volovich: STOCHASTIC LIMIT METHOD AND INTERFERENCE IN QUANTUM MANY–PARTICLE SYSTEMS Theoretical and Mathematical Physics, 183 (3) (2015) 782–799
32. Bellman R.: Adaptive control processes: A guided tour, Princeton University Press (1961)
33. Neill Lambert, Yueh-Nan Chen, Yuan-Chung Cheng, Che-Ming Li, Guang-Yin Chen, Franco Nori: Quantum biology NATURE PHYSICS 9, JANUARY (2013) 10-18 www.nature.com/naturephysics
34. Hisaharu Umegaki: Operator algebras and mathematical information theory, Kaigai Publications, LTD (1985)
35. J. von Neumann: Mathematical foundations of Quantum Mechanics, Princeton University Press (1955) also: Mathematische Grundlagen der Quantenmechanik, Berlin, Springer (1932), Dover–Usa (1943)

CONTENTS

Quantum Bio-Informatics VI
© 2020 World Scientific Publishing Co. Pte. Ltd.
pp. 1–28

CANONICAL FORMS OF QUADRATIC BOSON FIELDS

LUIGI ACCARDI

Centro Vito Volterra, Università di Roma Tor Vergata, Italy
E-mail: accardi@volterra.mat.uniroma2.it

ANDREAS BOUKAS

Centro Vito Volterra, Università di Roma Tor Vergata, Italy and Graduate
School of Mathematics, Hellenic Open University, Greece
E-mail: boukas.andreas@ac.eap.gr, andreasboukas@yahoo.com

1. Introduction

The present paper is a contribution to our program of calculating the vacuum distributions of the truncated Virasoro fields. In [2] these distributions were computed for a very special class of fields and were found to be products of Gamma distributions.

The truncated Virasoro fields can be realized as skew–adjoint quadratic expressions in usual Boson fields, these are called **quadratic fields**. A quadratic field can be written in several ways as a quadratic expression in first order boson fields, so in order to simplify calculations it is useful to look for **canonical forms** in which these expressions are as simple as possible. The most widespread of such forms consists in indexing the 1–st order operators not with general test functions, but with elements of an ortho–normal basis (c_j) of the 1–particle space so that, for fixed $d \leq \infty$ (in fact our main results deal with the finite dimensional case, but we used notations that suggest a natural transition to the infinite dimensional case when such a transition is possible) one is reduced quadratic expressions in the boson operators

$$[a_{e_i}, a_{e_j}^+] = \delta_{i,j} \quad ; \quad i, j \in \{1, \ldots, d\}$$

These are what we call **standard forms** (see sections 2 and 2.2). In order to use some existing theorems in linear algebra, we will have to consider also boson operators $a_{f_j}^\pm$ indexed by a general **linear basis** of the 1–particle space. This leads to study how the coefficients of a quadratic

1

field vary under a change of linear basis $a_{e_j}^{\pm} \mapsto a_{f_j}^{\pm}$ (see section 2.5). More generally, if G is a group acting on the 1–st order boson operators, one can consider the orbit of the coefficients of a quadratic field under this action. A **canonical form** is a special point in this orbit, usually selected because of the simplicity of its form. We discuss three canonical forms with respect to the action of the general linear group on the 1–particle space (see sections 3.4, 3.5 and 3.6).

In addition to the general linear group on the 1–particle space, we also consider the translation group (section 3.3) and we prove that, contrarily to the 1–dimensional case where any quadratic field is translation–equivalent to a purely homogeneous field (i.e. without first order terms), in dimensions ≥ 2 some obstructions appear. For example in the case of Virasoro fields, it is not yet clear if these conditions are satisfied.

We know from the 1–dimensional case that the set of vacuum distributions of quadratic fields can consist of strongly inequivalent probability measures, i.e. measures which are not obtained one from another by changing some numerical parameters within a single class. We conjecture that the canonical forms mentioned above are strictly related with the structure of the vacuum distributions of the elements of a given orbit. This is motivated by the result of [4] according to which the vacuum distributions of quadratic fields, diagonalizable in an ortho–normal basis, are product measures. So we expect that more complex canonical forms give rise to non–product measures.

2. Quadratic fields

Let $\mathcal{T}$ be a finite dimensional vector space (always complex unless otherwise specified) called the **test function space**. A **linear basis** (resp. **total linear basis**) of $\mathcal{T}$ is a sub–set $(e_j)_{j \in D} \subset \mathcal{T}$, D an at most countable set, such that for all $\xi \in \mathcal{T}$ there exists a finite set $I \subseteq D$ and $\xi_j \in \mathbb{C}$ $(j \in I)$ such that

$$\xi = \sum_{j \in I} \xi_j e_j \tag{1}$$

(resp. the space of ξ satisfying (1) is dense in $\mathcal{T}$). Let $\mathcal{P}$ be a $*$–algebra (always complex and with identity unless otherwise specified) and let

$$a^+ : \mathcal{T} \to \mathcal{P} \tag{2}$$

be a linear map. Define

$$a_f := (a_f^+)^* \quad ; \quad f \in \mathcal{T}$$

Without loss of generality one can suppose that

$$\mathcal{P} = \text{algebraic span of } \{a_f^+, a_g \; : \; f, g \in \mathcal{T}\} \tag{3}$$

In the following we suppose that (3) is satisfied.

Definition 2.1. A **quadratic field** is an Hermitean element of $\mathcal{P}$ of the form

$$X := \sum_{f,g \in F} \left(X_{1,f,g} a_f^+ a_g^+ + X_{2,f,g} a_f a_g^+ + X_{3,f,g} a_f^+ a_g + X_{4,f,g} a_f a_g \right) + c \tag{4}$$

$$+ \sum_{h \in G} Z_{1,h} a_h^+ + \sum_{h \in G} Z_{2,h} a_h + \mu$$

where $F \subseteq \mathcal{T}$ is a **finite set** and $c \in \mathbb{R}$. A **homogeneous quadratic field** is an Hermitean element of $\mathcal{P}$ of the form

$$X := \sum_{f,g \in F} \left(X_{1,f,g} a_f^+ a_g^+ + X_{2,f,g} a_f a_g^+ + X_{3,f,g} a_f^+ a_g + X_{4,f,g} a_f a_g \right) \tag{5}$$

where $F \subseteq \mathcal{T}$ is a finite set and $X_{1,f,g}, X_{1,f,g}, X_{2,f,g} = \bar{X}_{2,g,f} \in \mathbb{C}$.
If a quadratic field is a positive operator, then it is called a **quadratic Hamiltonian**. Two quadratic fields (13) are called **equivalent** if they define the same operator up to an additive real constant.

The right hand side of (5) is not unique: several different quadratic expressions can describe the same operator X. Some of these expressions however may be more convenient than others for concrete calculations.

Lemma 2.1. *Any homogeneous quadratic field X can be written in the form*

$$X = \sum_{f,g \in F} \left(Y_{1,f,g} a_f^+ a_g^+ + Y_{2,f,g} a_f a_g^+ + Y_{3,f,g} a_f^+ a_g + \bar{Y}_{1,g,f} a_f a_g \right) \tag{6}$$

where

$$\bar{Y}_{j,f,g} = Y_{j,g,f} \; ; \quad j = 2, 3 \tag{7}$$

Proof. By definition an homogeneous quadratic field X has the form (5). Using (8) one has (summation over repeated indexes is understood):

$$\left(X_{1,f,g} a_f^+ a_g^+ + X_{2,f,g} a_f a_g^+ + X_{3,f,g} a_f^+ a_g + X_{4,f,g} a_f a_g \right) = X$$

$$= X^* = \left(\bar{X}_{1,f,g} a_g a_f + \bar{X}_{2,f,g} a_g a_f^+ + \bar{X}_{3,f,g} a_g^+ a_f + \bar{X}_{4,f,g} a_g^+ a_f^+ \right)$$

$$= \left(\bar{X}_{1,g,f} a_f a_g + \bar{X}_{2,g,f} a_f a_g^+ + \bar{X}_{3,g,f} a_f^+ a_g + \bar{X}_{4,g,f} a_f^+ a_g^+ \right)$$

It follows that

$$X = \frac{1}{2}(X + X^*)$$

$$= \frac{1}{2} \left(X_{1,f,g} a_f^+ a_g^+ + X_{2,f,g} a_f a_g^+ + X_{3,f,g} a_f^+ a_g + X_{4,f,g} a_f a_g \right)$$

$$+ \frac{1}{2} \left(\bar{X}_{1,g,f} a_f a_g + \bar{X}_{2,g,f} a_f a_g^+ + \bar{X}_{3,g,f} a_f^+ a_g + \bar{X}_{4,g,f} a_f^+ a_g^+ \right)$$

$$= \frac{1}{2} \left(X_{1,f,g} + \bar{X}_{4,g,f} \right) a_f^+ a_g^+ + \frac{1}{2} \left(X_{2,f,g} + \bar{X}_{2,g,f} \right) a_f a_g^+$$

$$+ \frac{1}{2} \left(\bar{X}_{3,g,f} + X_{3,f,g} \right) a_f^+ a_g + \frac{1}{2} \left(\bar{X}_{1,g,f} + X_{4,f,g} \right) a_f a_g$$

Thus defining

$$Y_{1,f,g} := \frac{1}{2} \left(X_{1,f,g} + \bar{X}_{4,g,f} \right) \ ; \ Y_{2,f,g} := \frac{1}{2} \left(X_{2,f,g} + \bar{X}_{2,g,f} \right) \ ;$$

$$Y_{3,f,g} := \frac{1}{2} \left(X_{3,f,g} + \bar{X}_{3,g,f} \right)$$

one has

$$X = Y_{1,f,g} a_f^+ a_g^+ + Y_{2,f,g} a_f a_g^+ + Y_{3,f,g} a_f^+ a_g + \bar{Y}_{1,g,f} a_f a_g$$

which is (9). Moreover

$$\bar{Y}_{j,f,g} := \frac{1}{2} \left(X_{j,f,g} + \bar{X}_{j,g,f} \right)^-$$

$$= \frac{1}{2} \left(X_{j,g,f} + \bar{X}_{j,f,g} \right) = Y_{j,g,f} \ ; \quad j = 2,3$$

which is (7). $\square$

Corollary 2.1. *Suppose that the creators commute:*

$$a_f^+ a_g^+ = a_g^+ a_f^+ \tag{8}$$

Then any homogeneous quadratic field X can be written in the form

$$X = \sum_{f,g \in F} \left(Z_{1,f,g} a_f^+ a_g^+ + Y_{2,f,g} a_f a_g^+ + Y_{3,f,g} a_f^+ a_g + \bar{Z}_{1,f,g} a_f a_g \right) \tag{9}$$

where $Y_{2,f,g}$ and $Y_{3,f,g}$ satisfy (7) and

$$Z_{1,f,g} = Z_{1,g,f} \qquad (10)$$

Proof. From Lemma 2.1 we know that X has the form (9). The commutativity of creators then implies that

$$\sum_{f,g\in F} Y_{fg} a_f^+ a_g^+ = \sum_f Y_{ff} a_f^+ a_f^+ + \sum_{f\neq g} Y_{fg} a_f^+ a_g^+$$

$$= \sum_f Y_{ff} a_f^+ a_f^+ + \frac{1}{2}\sum_{f\neq g} Y_{fg} a_f^+ a_g^+ + \frac{1}{2}\sum_{f\neq g} Y_{gf} a_g^+ a_f^+$$

$$= \sum_f Y_{ff} a_f^+ a_f^+ + \sum_{f\neq g} \frac{1}{2}(Y_{fg} + Y_{gf}) a_f^+ a_g^+ \qquad (11)$$

Thus, defining for any $f,g \in F$

$$Z_{fg} := \frac{1}{2}(Y_{fg} + Y_{gf}) \qquad (12)$$

(12) implies that (10) holds and, taking the adjoint of (11), one deduces (9). $\square$

2.1. Matrix notations for homogeneous quadratic fields

From now on we will suppose that the creators commute. We enumerate the set F in the sum in (9) obtaining the identification

$$F = \{f_1, \ldots, f_N\}$$

Introducing the notation

$$(a_F^+ \ a_F) := \left(a_{f_1}^+ \ \cdots \ a_{f_N}^+ \ a_{f_1} \ \cdots \ a_{f_N} \right)$$

one can write the homogeneous quadratic field (9) in matrix form as follows:

$$X = (a_F^+ \ a_F) \begin{pmatrix} X_{1,F} & X_{2,F} \\ X_{3,F} & \bar{X}_{1,F} \end{pmatrix} \begin{pmatrix} a_F^+ \\ a_F \end{pmatrix}$$

$$= \begin{pmatrix} a_{f_1}^+ & \cdots & a_{f_N}^+ & a_{f_1} & \cdots & a_{f_N} \end{pmatrix}$$

$$\times \begin{pmatrix} X_{1,f_1,f_1} & \cdots & X_{1,f_1,f_N} & X_{2,f_1,f_1} & \cdots & X_{2,f_1,f_N} \\ \vdots & \vdots & \vdots & \vdots & \vdots & \vdots \\ X_{1,f_N,f_1} & \cdots & X_{1,f_N,f_N} & X_{2,f_N,f_1} & \cdots & X_{2,f_N,f_N} \\ X_{3,f_1,f_1} & \cdots & X_{3,f_1,f_N} & \bar{X}_{1,f_1,f_1} & \cdots & \bar{X}_{1,f_1,f_N} \\ \vdots & \vdots & \vdots & \vdots & \vdots & \vdots \\ X_{3,f_N,f_1} & \cdots & X_{3,f_N,f_N} & \bar{X}_{1,f_N,f_1} & \cdots & \bar{X}_{1,f_N,f_N} \end{pmatrix} \begin{pmatrix} (a_{f_1}^+ \\ \vdots \\ a_{f_N}^+ \\ a_{f_1} \\ \vdots \\ a_{f_N} \end{pmatrix}$$

In particular, up to equivalence, one can suppose that X has the form:

$$X = \sum_{f,g \in F} (a_f^+, a_f) \begin{pmatrix} X_{1,f,g} & X_{3,f,g} \\ X_{2,f,g} & \bar{X}_{1,f,g} \end{pmatrix} \begin{pmatrix} a_g^+ \\ a_g \end{pmatrix} \tag{13}$$

$$= \sum_{f,g \in F} (a_f^+ X_{1,f,g} + a_f X_{2,f,g} \, , \, a_f^+ X_{3,f,g} + a_f \bar{X}_{1,f,g}) \begin{pmatrix} a_g^+ \\ a_g \end{pmatrix}$$

$$:= \sum_{f,g \in F} \left(X_{1,f,g} a_f^+ a_g^+ + X_{2,f,g} a_f a_g^+ + X_{3,f,g} a_f^+ a_g + \bar{X}_{1,f,g} a_f a_g \right)$$

2.2. *Boson quadratic fields*

From now on $\mathcal{T}$ is a Hilbert space with scalar product $\langle \cdot \, , \, \cdot \rangle$. If $(e_j)_{j \in D} \subset \mathcal{T}$ is a linear basis (resp. total linear basis) of $\mathcal{T}$ and the $(e_j)_{j \in D}$ are mutually orthogonal and normalized we speak of an **orthonormal linear basis** (resp. **orthonormal basis**) of $\mathcal{T}$.

Denote $\Gamma(\mathcal{T})$ the boson Fock space over $\mathcal{T}$, Φ its vacuum vector and $\mathcal{P} \cdot \Phi$ the domain of number vectors. $\mathcal{P} \cdot \Phi$ is invariant under the action of the $*$–algebra generated by the creation and annihilation operators $a_f^{\pm}$ ($f \in \mathcal{T}$).

Let (e_j) be a linear (or total linear) basis of $\mathcal{T}$. If $\hat{A}$ is a linear operator on $\mathcal{T}$, to the pair $(\hat{A}, (e_j))$ one associates the matrix $A^{(e)} := (A_{kj}^{(e)})$ defined by:

$$A^{(e)} e_j = \sum_k A_{kj}^{(e)} e_k \tag{14}$$

This is motivated by the fact that, if $f = \sum_j f_j e_j \in \mathcal{T}$, then one has

$$A^{(e)} f = \sum_j f_j A^{(e)} e_j = \sum_j f_j \sum_k A_{kj}^{(e)} e_k \tag{15}$$

$$= \sum_k (\sum_j A_{kj}^{(e)} f_j) e_k \tag{16}$$

Thus, in the identification of $\mathcal{T}$ with $\mathbb{C}^d$ given by the basis (e_j), the action of $\hat{A}$ on vectors will be given by the matrix $(A^{(e)})^T$. If $\mathcal{T}$ is endowed with a semi–scalar product $\langle \, \cdot \, , \, \cdot \, \rangle$, and (e_j) is an orthonormal basis, then

$$A^{(e)}_{kj} = \langle e_k, A^{(e)} e_j \rangle$$

Denote $M_d(\mathbb{C})$ the algebra of $d \times d$ matrices with complex entries. For $f \equiv (f_{ij}) \in M_d(\mathbb{C})$,

$$f^* \equiv ((f^*)_{ij}) := (\bar{f}_{ji}) \qquad ; \qquad \forall f := (f_{ij}) \in M_d(\mathbb{C}) \qquad (17)$$

denotes the **adjoint** of f and

$$(f^T)_{ij} := f_{ji} \qquad (18)$$

its **transpose**. $f \in M_d(\mathbb{C})$ is called **self–adjoint** if $f = f^*$, **symmetric** if $f = f^T$. $M_{d,sym}(\mathbb{C})$ denotes the space of symmetric matrices

$$f_{hk} = f_{kh}$$

i.e. the range of the linear projector

$$f = (f_{hk}) \in M_d(\mathbb{C}) \;\mapsto\; (f_s)_{hk} := \frac{1}{2}(f_{hk} + f_{kh}) \qquad (19)$$

Let (e_j) be a linear basis of $\mathcal{T}$. For the matrix elements (or coefficients) of a linear operator A in the (e_j)–basis, we use the convention:

$$Ae_j = \sum_k A_{kj} e_k \qquad (20)$$

Then, if $f = \sum_j f_j e_j$, the matrix (A_{ij}) acts on the coordinates (f_i) in the usual way:

$$Af = \sum_j f_j Ae_j = \sum_j f_j \sum_k A_{kj} e_k \qquad (21)$$

$$= \sum_k (\sum_j A_{kj} f_j) e_k \qquad (22)$$

Remark. Notice that, for a symmetric operator A, one has

$$\bar{A} = A^*$$

where $\bar{A}$ denotes entrywise complex conjugate. In fact

$$(A^*)_{ij} = \bar{A}_{ji} = \bar{A}_{ij} = (\bar{A})_{ij}$$

In particular, if the matrix A has real entries, then it is Hermitean because

$$A_{ij} = \bar{A}_{ij} = \bar{A}_{ji} = A_{ji}$$

8

Lemma 2.2. *For any linear basis (e_j) of $\mathcal{T}$:*
(i) The set $\{a_{e_j}a_{e_h} : j \leq h \in D\}$ is linearly independent.
(ii) The set $\{a_{e_j}^+ a_{e_h} : j, h \in D\}$ is linearly independent.

Remark. Since the annihilators commute, the set $\{a_{e_j}a_{e_h} : j, k \in D\}$ cannot be linearly independent.

Proof. Suppose that

$$\sum_{j\leq h} x_{jh} a_{e_j} a_{e_h} = 0$$

Then for any e_k in the given basis, one has

$$0 = \sum_{j\leq h} x_{jh}[a_{e_k}^+, a_{e_j} a_{e_h}] = \sum_{j\leq h} x_{jh}[a_{e_k}^+, a_{e_j}]a_{e_h}$$
$$+ \sum_{j\leq h} x_{jh} a_{e_j}[a_{e_k}^+, a_{e_h}]$$

$$= \sum_{j\leq h} x_{jh}\langle e_k, e_j\rangle a_{e_h} + \sum_{j\leq h} x_{jh} a_{e_j}\langle e_k, e_h\rangle$$

Therefore, for any $h, k \in D$,

$$0 = \sum_{j\leq h} x_{jh}\langle e_k, e_j\rangle[a_{e_l}^+, a_{e_h}] + \sum_{j\leq h} x_{jh}[a_{e_l}^+, a_{e_j}]\langle e_k, e_h\rangle$$

$$= \sum_{j\leq h} x_{jh}\langle e_k, e_j\rangle\langle e_l, e_h\rangle + \sum_{j\leq h} x_{jh}\langle e_l, e_j\rangle\langle e_k, e_h\rangle$$

$$= \sum_{j\leq h} \langle e_k, e_j\rangle x_{jh}\overline{\langle e_h, e_l\rangle} + \sum_{j\leq h} \langle e_l, e_j\rangle x_{jh}\overline{\langle e_h, e_k\rangle}$$

Then, defining $x_{jh} := 0$ for $j > h$, $x := (x_{jh})$, $S := (S_{kh}) = (\langle e_k, e_h\rangle)$ one has

$$0 = (Sx\bar{S})_{kl} + (Sx\bar{S})_{lk} \iff 0 = Sx\bar{S} + (Sx\bar{S})^T = Sx\bar{S} + \bar{S}^T x^T S^T$$

and from

$$\bar{S}_{hm} := \langle e_m, e_h\rangle = S_{mh} = (S^T)_{hm} \iff \bar{S} = S^T \iff S^* = S$$

it follows that

$$0 = SxS^T + Sx^T S^T \iff 0 = x + x^T$$

because S is non–singular. Since x is upper–triangular, this is equivalent to say that $x = 0$. Now suppose that

$$\sum_{j,h} x_{jh} a_{e_j}^+ a_{e_h}^- = 0$$

Then for any e_n in the given basis, one has

$$0 = \sum_{j,h} x_{jh} a_{e_j}^+ [a_{e_n}^+, a_{e_h}^-] = -\sum_{j,h} x_{jh} a_{e_j}^+ \langle e_n, e_h \rangle$$

Therefore, for any $m, n \in D$,

$$0 = \sum_{j,h} x_{jh} [a_{e_m}, a_{e_j}^+] \langle e_n, e_h \rangle = \sum_{j,h} \langle e_m, e_j \rangle x_{jh} \langle e_n, e_h \rangle = (Sx\bar{S})_{mn}$$

thus $Sx\bar{S} = 0$ and since S, hence $\bar{S}$, is invertible this is equivalent to $x = 0$.
☐

2.3. Standard forms of homogeneous boson quadratic fields

Proposition 2.1. *For any linear basis (e_j) of T and for any homogeneous quadratic field X there exist matrices $A := (A_{ij})$ and $C := (C_{ij})$ such that X is equivalent to a field of the form:*

$$\sum_{i,j} \left(A_{ij} a_{e_i}^+ a_{e_j}^+ + \bar{A}_{ji} a_{e_j} a_{e_i} + C_{ij} a_{e_i}^+ a_{e_j} \right) \tag{23}$$

with additive constant given by

$$c_X := \sum_{f,g \in F} Z_{fg} \langle f, g \rangle \tag{24}$$

Moreover the matrix $A = (A_{ij})$ can be taken to be symmetric and the matrix $C = (C_{ij})$ is Hermitean:

$$A_{ij} = A_{ji} \quad ; \quad \bar{C}_{ij} = C_{ji} \quad ; \quad \forall i, j \in D \tag{25}$$

Finally, for fixed (e_j), the matrices A and C with the above properties are uniquely determined.

Proof. Due to the identity

$$\sum_{f,g \in F} Z_{fg} a_f^- a_g^+ = \sum_{f,g \in F} Z_{fg} [a_f^-, a_g^+] + \sum_{f,g \in F} Z_{fg} a_g^+ a_f^-$$

$$= \sum_{f,g \in F} Z_{fg} a_g^+ a_f^- + \sum_{f,g \in F} Z_{fg} \langle f, g \rangle$$

$$= \sum_{f,g \in F} Z_{gf} a_f^+ a_g^- + \sum_{f,g \in F} Z_{fg} \langle f, g \rangle$$

any quadratic field F (49) is equivalent to a quadratic field of the form

$$X = \sum_{f,g \in F} \left(X_{fg} a_f^+ a_g^+ + \bar{X}_{gf} a_g a_f + (Y_{fg} + Z_{gf}) a_f^+ a_g \right)$$

with additive constant given by (24). For (e_j) as in the statement and for any $f \in F$, let $f = \sum_j f_j e_j$ be the expansion of f in this basis. Then, renaming $W_{fg} + Z_{gf} \to Z_{fg}$, up to strict equivalence one has

$$F = \sum_{f,g \in F} \left(X_{fg} a_{\sum_j f_j e_j}^+ a_{\sum_h g_h e_h}^+ + Y_{gf} a_{\sum_h g_h e_h} a_{\sum_j f_j e_j} \right.$$

$$\left. + Z_{fg} a_f^+ a_{\sum_h g_h e_h} \right)$$

$$= \sum_{f,g \in F} X_{fg} \sum_{j,h} f_j g_h a_{e_j}^+ a_{e_h}^+ + \sum_{f,g \in F} Y_{gf} \sum_{j,h} \bar{f}_j \bar{g}_h a_{e_h} a_{e_j}$$

$$+ \sum_{f,g \in F} Z_{fg} \sum_{j,h} f_j \bar{g}_h a_{e_j}^+ a_{e_h}$$

$$= \sum_{j,h} \left(\sum_{f,g \in F} (X_{fg} f_j g_h) a_{e_j}^+ a_{e_h}^+ + \sum_{f,g \in F} (Y_{gf} \bar{f}_j \bar{g}_h) a_{e_h} a_{e_j} \right.$$

$$\left. + \sum_{f,g \in F} (Z_{fg} f_j \bar{g}_h) a_{e_j}^+ a_{e_h} \right)$$

$$= \sum_{j,h} \left(\sum_{f,g \in F} (X_{fg} f_j g_h) a_{e_j}^+ a_{e_h}^+ + \sum_{f,g \in F} (Y_{fg} \bar{f}_j \bar{g}_h) a_{e_j} a_{e_h} \right.$$

$$\left. \sum_{f,g \in F} (Z_{fg} f_j \bar{g}_h) a_{e_j}^+ a_{e_h} \right)$$

Thus defining

$$A_{jh} := \sum_{f,g \in F} X_{fg} f_j g_h \quad ; \quad B_{jh} := Y_{fg} f_j g_h \quad ; \quad C_{jh} := \sum_{f,g \in F} Z_{fg} f_j \bar{g}_h$$

one obtains

$$F = \sum_{j,h} \left(A_{jh} a_{e_j}^+ a_{e_h}^+ + B_{jh} a_{e_j} a_{e_h} + C_{jh} a_{e_j}^+ a_{e_h} \right) \tag{26}$$

By definition one has $F = F^*$, i.e.

$$\sum_{j,h} \left(A_{jh} a_{e_j}^+ a_{e_h}^+ + B_{jh} a_{e_j} a_{e_h} + C_{jh} a_{e_j}^+ a_{e_h} \right)$$
$$= \sum_{j,h} \left(\bar{A}_{jh} a_{e_h}^- a_{e_j}^- + \bar{B}_{jh} a_{e_h}^+ a_{e_j}^+ + \bar{C}_{jh} a_{e_h}^+ a_{e_j}^- \right)$$

Since the ranges of the $a_{e_h}^+ a_{e_j}^+$, $a_{e_h}^- a_{e_j}^-$, $a_{e_h}^+ a_{e_j}^-$ are mutually orthogonal and the creators (resp. annihilators) commute, it follows that

$$\sum_{j,h} B_{jh} a_{e_j} a_{e_h} = \sum_{j,h} \bar{A}_{jh} a_{e_h}^- a_{e_j}^- = \sum_{j,h} \bar{A}_{hj} a_{e_j}^- a_{e_h}^- \qquad (27)$$

$$\sum_{j,h} C_{jh} a_{e_j}^+ a_{e_h} = \sum_{j,h} \bar{C}_{jh} a_{e_h}^+ a_{e_j}^- = \sum_{j,h} \bar{C}_{hj} a_{e_j}^+ a_{e_h}^- \qquad (28)$$

(27) implies that in (26) one can replace $\sum_{j,h} B_{jh} a_{e_j} a_{e_h}$ by $\sum_{j,h} \bar{A}_{hj} a_{e_j}^- a_{e_h}^-$ thus finding

$$F = \sum_{j,h} \left(A_{jh} a_{e_j}^+ a_{e_h}^+ + \bar{A}_{hj} a_{e_j}^- a_{e_h}^- + C_{jh} a_{e_j}^+ a_{e_h} \right) \qquad (29)$$

Moreover, due to the commutativity of creators, one has

$$\sum_{i,j} A_{ij} a_i^+ a_j^+ = \sum_{i,j} A_{ij} a_j^+ a_i^+ = \sum_{i,j} A_{ji} a_i^+ a_j^+$$

so that

$$\sum_{i,j} A_{ij} a_i^+ a_j^+ = \frac{1}{2} \left(\sum_{i,j} A_{ij} a_i^+ a_j^+ + \sum_{i,j} A_{ji} a_i^+ a_j^+ \right)$$
$$= \sum_{i,j} \frac{A_{ij} + A_{ji}}{2} a_i^+ a_j^+$$

Therefore in (29) we can suppose, up to equivalence, that A is symmetric and this implies (23). From (28) and the linear independence of the $a_h^+ a_j$ one deduces the second identity in (25). Finally, if $A' = (A_{ij})$ and $C' = (C_{ij})$ is another pair of matrices satisfying the conclusions of the theorem, the orthogonality of the ranges of the $a_{e_h}^+ a_{e_j}^+$, $a_{e_h}^- a_{e_j}^-$, $a_{e_h}^+ a_{e_j}^-$ implies that

$$\sum_{j,h} \bar{A}_{jh} a_{e_h}^- a_{e_j}^- = \sum_{j \leq h} \bar{A}_{jh} a_{e_h}^- a_{e_j}^- = \sum_{j \leq h} \bar{A}'_{jh} a_{e_h}^- a_{e_j}^-$$
$$= \sum_{j,h} \bar{A}'_{jh} a_{e_h}^- a_{e_j}^-$$

$$\sum_{j,h} C_{jh} a^+_{e_j} a_{e_h} = \sum_{j,h} C'_{jh} a^+_{e_j} a_{e_h}$$

and linear independence implies that $A = A'$, $C = C'$.

Definition 2.2. The expression (23), with $A := (A_{ij})$ and $C := (C_{ij})$ satisfying (25) will be called the **standard form** of the homogeneous quadratic field F with respect to the (e_j)–basis.

2.4. The $*$–Lie algebra of homogeneous quadratic fields

From now on, unless explicitly stated, we fix a basis (e_j) and we study the $*$–Lie algebra generated by the homogeneous quadratic fields in their standard form with respect to this basis.

Proposition 2.1 shows that the generators of this $*$–Lie algebra are in one–to–one correspondence with the pairs of matrices

$$(A, C) \in M_{d,sym}(\mathbb{C}) \times M_{d,herm}(\mathbb{C})$$

with A symmetric and C Hermitean. In this section we work with a fixed linear basis and we omit it from notations.

Lemma 2.3. *With the convention of summation over repeated indices, for $A \equiv (A_{ij}) \in M_{d,sym}(\mathbb{C})$ define*

$$B_0^2(A) := A_{hk} a^+_h a^+_k := \sum_{h,k \in \{1,\ldots,d\}} A_{hk} a^+_h a^+_k \tag{30}$$

$$B_2^0(A) := (B_0^2(A))^* = \sum_{h,k \in \{1,\ldots,d\}} \bar{A}_{hk} a^-_h a^-_k \tag{31}$$

For $C \equiv (C_{ij}) \in M_{d,herm}(\mathbb{C})$ define

$$B_1^1(f) := f_{ij} a^+_i a_j =: \sum_{i,j \in \{1,\ldots,d\}} f_{ij} a^+_i a_j \tag{32}$$

Then, recalling from (19) that M_s denotes the symmetric part of the matrix $M \in M_d(\mathbb{C})$, one has:

$$B_0^2(M) = B_0^2(M_s) \tag{33}$$

$$B_2^0(M) = B_2^0(M^*) = B_2^0(\overline{M_s}) \tag{34}$$

$$(B_1^1(M))^* = B_1^1(M^*) \tag{35}$$

Proof. The commutativity of creators implies that

$$B_0^2(M) = M_{hk}a_h^+ a_k^+ = M_{kh}a_h^+ a_k^+$$
$$= \frac{1}{2}(M_{hk} + M_{kh})a_h^+ a_k^+ = B_0^2(M_s) \quad ; \quad \forall f \in M_d(\mathbb{C})$$

which is (33). (34) follows from:

$$B_2^0(M) = (B_0^2(M))^* = (M_{ij}a_i^+ a_j^+)^* = \overline{M_{ij}}a_j a_i$$
$$= \overline{M_{ji}}a_i a_j = (M^*)_{ij}a_i a_j = (\bar{M}_s)_{ij}a_i a_j$$

Then one has

$$(B_1^1(M))^* = (M_{ij}a_i^+ a_j)^* = \overline{M_{ij}}a_j^+ a_i$$
$$= \overline{M_{ji}}a_i^+ a_j = a_i^+ (M^*)_{ij}a_j = B_1^1(M^*)$$

Lemma 2.4. *The following commutation relations take place for any* $M, N \in M_d(\mathbb{C})$ *and* $A_1, A_2 \in M_{d,sym}(\mathbb{C})$:

$$[B_2^0(A_1), B_0^2(A_2)] = 2\,Tr(A_1^* A_2) + 4B_1^1(A_2 A_1^*) \tag{36}$$

$$[B_1^1(M), B_0^2(A)] = B_0^2((MA) + (MA)^T) \tag{37}$$

$$[B_1^1(M), B_2^0(A)] = -B_2^0(AM + (AM)^T) \tag{38}$$

$$[B_1^1(M), B_1^1(N)] = B_1^1([M, N]) \tag{39}$$

Proof. One has:

$$[B_2^0(A_1), B_0^2(g)] = \overline{A_{1;ij}}A_{2;hk}[a_i a_j, a_h^+ a_k^+]$$
$$= \overline{A_{1;ij}}A_{2;hk}\left([a_i a_j, a_h^+]a_k^+ + a_h^+[a_i a_j, a_k^+]\right)$$

$$= (A_1)_{ji}^* A_{2;hk}\left(a_i[a_j, a_h^+]a_k^+ + [a_i, a_h^+]a_j a_k^+ \right.$$
$$\left. + a_h^+ a_i[a_j, a_k^+] + a_h^+[a_i, a_k^+]a_j\right)$$

$$= (A_1)_{ji}^* A_{2;hk}\left(a_i[a_j, a_h^+]a_k^+ + [a_i, a_h^+]a_j a_k^+\right.$$
$$\left. a_h^+ a_i[a_j, a_k^+] + a_h^+[a_i, a_k^+]a_j\right)$$

$$= (A_1)^*_{ji} A_{2;hk} \left(a_i a_k^+ \delta_{jh} + a_j a_k^+ \delta_{ih} + a_h^+ a_i \delta_{jk} + a_h^+ a_j \delta_{ik} \right)$$

$$= (A_1)^*_{ji} A_{2;hk} \left(\delta_{ik} \delta_{jh} + a_k^+ a_i \delta_{jh} + \delta_{jk} \delta_{ih} \right.$$
$$\left. a_k^+ a_j \delta_{ih} + a_h^+ a_i \delta_{jk} + a_h^+ a_j \delta_{ik} \right)$$

$$= (A_1)^*_{ji} A_{2;hk} \delta_{ik} \delta_{jh} + (A_1)^*_{ji} A_{2;hk} a_k^+ a_i \delta_{jh} + (A_1)^*_{ji} A_{2;hk} \delta_{jk} \delta_{ih}$$
$$+ (A_1)^*_{ji} A_{2;hk} a_k^+ a_j \delta_{ih} + (A_1)^*_{ji} A_{2;hk} a_h^+ a_i \delta_{jk} + (A_1)^*_{ji} A_{2;hk} a_h^+ a_j \delta_{ik}$$

$$= (A_1)^*_{ji} A_{2;ji} + (A_1)^*_{ji} A_{2;jk} a_k^+ a_i + (A_1)^*_{ji} A_{2;ij}$$
$$+ (A_1)^*_{ji} A_{2;ik} a_k^+ a_j + (A_1)^*_{ji} A_{2;hj} a_h^+ a_i + (A_1)^*_{ji} A_{2;hi} a_h^+ a_j$$

Using the symmetry of A_2 this becomes

$$= 2(A_1)^*_{ji} A_{2;ij} + A_{2;kj} (A_1)^*_{ji} a_k^+ a_i + (A_1)^*_{ji} A_{2;ik} a_k^+ a_j$$
$$+ A_{2;hj} (A_1)^*_{ji} a_h^+ a_i + (A_1)^*_{ji} A_{2;ih} a_h^+ a_j$$

$$= 2\mathrm{Tr}(A_1^* A_2) + (A_2 A_1^*)_{ki} a_k^+ a_i + (A_1^* A_2)_{jk} a_k^+ a_j$$
$$+ (A_2 A_1^*)_{hi} a_h^+ a_i + (A_1^* A_2)_{jh} a_h^+ a_j$$

$$= 2\mathrm{Tr}(A_1^* A_2) + 2(A_2 A_1^*)_{ki} a_k^+ a_i + 2(A_1^* A_2)_{ik} a_k^+ a_i$$

$$= 2\mathrm{Tr}(A_1^* A_2) + 2 \left((A_2 A_1^*)_{ki} + ((A_1^* A_2)^T)_{ki} \right) a_k^+ a_i$$

$$= 2\mathrm{Tr}(A_1^* A_2) + 2 \left((A_2 A_1^*)_{ki} + (A_2 (A_1^*)^T)_{ki} \right) a_k^+ a_i$$

$$= 2\mathrm{Tr}(A_1^* A_2) + 2 \left((A_2 A_1^*)_{ki} + (A_2 (A_1^T)^*)_{ki} \right) a_k^+ a_i$$

$$= 2\mathrm{Tr}(A_1^* A_2) + 2 \left((A_2 A_1^*)_{ki} + (A_2 A_1^*)_{ki} \right) a_k^+ a_i$$

$$= 2\mathrm{Tr}(A_1^* A_2) + 4 \left(A_2 A_1^* \right)_{ki} a_k^+ a_i$$

This proves (36). Similarly

$$[B_1^1(A_1), B_0^2(A_2)] = [A_{1;ij} a_i^+ a_j \ , \ A_{2;hk} a_h^+ a_k^+]$$
$$= A_{1;ij} A_{2;hk} [a_i^+ a_j \ , \ a_h^+ a_k^+]$$

$$= A_{1;ij} A_{2;hk} \left([a_i^+ a_j \ , \ a_h^+] a_k^+ + a_h^+ [a_i^+ a_j \ , \ a_k^+] \right)$$
$$= A_{1;ij} A_{2;hk} \left(a_i^+ [a_j \ , \ a_h^+] a_k^+ + a_h^+ a_i^+ [a_j \ , \ a_k^+] \right)$$

$$= A_{1;ij}A_{2;hk}a_i^+ a_k^+ \delta_{jh} + A_{1;ij}A_{2;hk}a_h^+ a_i^+ \delta_{jk}$$
$$= A_{1;ij}A_{2;jk}a_i^+ a_k^+ + A_{1;ij}A_{2;hj}a_h^+ a_i^+$$

$$= A_{1;ij}A_{2;jk}a_k^+ + A_{1;ij}A_{2;jh}a_h^+ a_i^+$$
$$= a_i^+ (A_1 A_2)_{ik}a_k^+ + a_h^+((A_1 A_2)^T)_{hi}a_i^+$$

$$= a_i^+((A_1 A_2) + (A_1 A_2)^T)_{ik}a_k^+ = B_0^2((A_1 A_2) + (A_1 A_2)^T)$$

which is (37). Taking the adjoint of this one obtains

$$[B_1^1(M), B_0^2(A)]^* = [B_0^2(A)^*, B_1^1(M)^*]$$
$$= [B_2^0(A^*), B_1^1(M^*)] = -[B_1^1(M^*), B_2^0(A^*)]$$

On the other hand

$$[B_1^1(M), B_0^2(A)]^* = B_0^2(MA + A^T M^T)^*$$

$$= B_0^2(A^* M^* + (M^T)^*(A^T)^*) = B_0^2(A^* M^* + (M^*)^T (A^*)^T)$$
$$= B_0^2(A^* M^* + (A^* M^*)^T)$$

Therefore

$$[B_1^1(M^*), B_2^0(A^*)] = -B_0^2(A^* M^* + (A^* M^*)^T)$$

Since M and A are arbitrary, one can replace them by M^* and A^* respectively, thus finding

$$[B_1^1(M), B_2^0(A)] = -B_0^2(AM + (AM)^T)$$

which is (38). Finally

$$[B_1^1(M), B_1^1(N)] = [M_{ij}a_i^+ a_j , N_{hk}a_h^+ a_k]$$
$$= M_{ij}N_{hk}[a_i^+ a_j , a_h^+ a_k]$$

$$- M_{lj}N_{hk}[a_i^+ a_j , a_h^+]a_k + M_{ij}N_{hk}a_h^+[a_i^+ a_j , a_k]$$

$$= M_{ij}N_{hk}a_i^+[a_j , a_h^+]a_k + M_{ij}N_{hk}a_h^+[a_i^+ , a_k]a_j$$

$$= M_{ij}N_{hk}a_i^+ \delta_{jh}a_k - M_{ij}N_{hk}a_h^+ \delta_{ki}a_j$$
$$= a_i^+ M_{ij}N_{jk}a_k - a_h^+ N_{hi}M_{ij}a_j$$

$$= a_i^+(MN)_{ik}a_k - a_h^+(NM)_{hj}a_j = B_1^1(MN) - B_1^1(NM)$$
$$= B_1^1(MN - NM) = B_1^1([M, N])$$

which is (39). $\square$

Lemma 2.5. *The following commutation relations take place for any* $C_1, C_2 \in M_{d,herm}(\mathbb{C})$ *and* $A_1, A_2 \in M_{d,sym}(\mathbb{C})$

$$[B_0^2(A_1) + B_1^1(C_1) + B_2^0(A_1) , \ B_0^2(A_2) + B_1^1(C_2) + B_2^0(A_2)] =$$

$$= 4B_1^1\left(A_2 A_1^* - (A_2 A_1^*)^* + [C_1, C_2]\right)$$

$$+ B_0^2((C_1 A_2 + (C_1 A_2)^T) - (C_2 A_1 + (C_2 A_1)^T)$$

$$B_2^0((A_1 C_2 + (A_1 C_2)^T) - (A_2 C_1 + (A_2 C_1)^T)$$

$$+ 2\, Tr(A_1^* A_2 - (A_1^* A_2)^*)$$

Proof. Using the commutation relations (36), (37), (38), (39) one finds

$$[B_0^2(A_1) + B_1^1(C_1) + B_2^0(A_1) , \ B_0^2(A_2) + B_1^1(C_2) + B_2^0(A_2)] =$$

$$= [B_0^2(A_1) , \ B_0^2(A_2)] + [B_0^2(A_1) , \ B_1^1(C_2)]$$
$$+ [B_0^2(A_1) , \ B_2^0(A_2)]$$

$$+ [B_1^1(C_1) , \ B_0^2(A_2)] + [B_1^1(C_1) , \ B_1^1(C_2)]$$
$$+ [B_1^1(C_1) , \ B_2^0(A_2)]$$

$$+ [B_2^0(A_1) , \ B_0^2(A_2)] + [B_2^0(A_1) , \ B_1^1(C_2)]$$
$$+ [B_2^0(A_1) , \ B_2^0(A_2)]$$

$$= [B_0^2(A_1) , \ B_1^1(C_2)] + [B_0^2(A_1) , \ B_2^0(A_2)]$$

$$+ [B_1^1(C_1) , \ B_0^2(A_2)] + [B_1^1(C_1) , \ B_1^1(C_2)]$$
$$+ [B_1^1(C_1) , \ B_2^0(A_2)]$$

$$+ [B_2^0(A_1) , \ B_0^2(A_2)] + [B_2^0(A_1) , \ B_1^1(C_2)]$$

$$= -B_0^2((C_2 A_1) + (C_2 A_1)^T)$$
$$- 2\mathrm{Tr}(A_2^* A_1) - 4B_1^1(A_1 A_2^*)$$

$$+B_0^2((C_1A_2) + (C_1A_2)^T) + B_1^1([C_1, C_2])$$
$$-B_2^0(A_2C_1 + (A_2C_1)^T)$$

$$2\mathrm{Tr}(A_1^*A_2) + 4B_1^1(A_2A_1^*) + B_2^0(A_1C_2 + (A_1C_2)^T)$$

$$= 4B_1^1(A_2A_1^*) + B_1^1([C_1, C_2]) - 4B_1^1(A_1A_2^*)$$

$$+B_0^2((C_1A_2) + (C_1A_2)^T) - B_0^2((C_2A_1) + (C_2A_1)^T)$$

$$B_2^0(A_1C_2 + (A_1C_2)^T) - B_2^0(A_2C_1 + (A_2C_1)^T)$$

$$+2\mathrm{Tr}(A_1^*A_2) - 2\mathrm{Tr}(A_2^*A_1)$$

$$= 4B_1^1\left(A_2A_1^* - (A_2A_1^*)^* + [C_1, C_2]\right)$$

$$+B_0^2((C_1A_2 + (C_1A_2)^T) - (C_2A_1 + (C_2A_1)^T))$$

$$B_2^0((A_1C_2 + (A_1C_2)^T) - (A_2C_1 + (A_2C_1)^T))$$

$$+2\mathrm{Tr}(A_1^*A_2 - (A_1^*A_2)^*)$$

$\square$

2.5. *Action of $GL(\mathcal{T})$ on standard forms of homogeneous quadratic boson fields*

In this section we fix $D := \{1, \ldots, d\}$ with $d \in \mathbb{N}^*$. In this case the standard form of an homogeneous quadratic boson field X with respect to a linear basis $c = (c_j)_{j \in D}$ of $\mathcal{T}$ can always be written as

$$X = \sum_{i,j \in D} \left(A_{ij}^{(e)} a_{e_i}^+ a_{e_j}^+ + \bar{A}_{ji}^{(e)} a_{e_j} a_{e_i} + C_{ij}^{(e)} a_{e_i}^+ a_{e_j} \right) \tag{40}$$

possibly introducing some coefficients $A_{ij}^{(e)}$ or $C_{ij}^{(e)}$ equal to zero. The pair of $d \times d$ matrices $(A^{(e)}, C^{(e)}) \equiv ((A_{ij}^{(e)}), (C_{ij}^{(e)}))$ and, under a linear change of basis, the field X remains the same, while the pair of matrices changes. Linear changes of bases are parametrized by the group $GL(\mathcal{T})$ of linear invertible transformations of $\mathcal{T}$. Thus these changes of bases induce an action of $GL(\mathcal{T})$ on the pairs $(A^{(e)}, C^{(e)})$ where $A^{(e)}$ is symmetric and $C^{(e)}$ Hermitean. The following Lemma determines the form of this action.

Lemma 2.6. *Let X be the homogeneous quadratic boson field given by (40). If $f \equiv (f_j)$ is another linear basis and $U := (U_{jk})$ is the coefficient matrix of e_k in the (f_j)–basis*

$$e_k := \sum_j U_{jk} f_j = U f_k \qquad ; \qquad U := (U_{jk}) \in GL(\mathbb{C}^d)$$

Then the standard form of X with respect to the linear basis $f \equiv (f_j)$ is characterized by the pair $(A^{(f)}, C^{(f)})$ given by:

$$C^{(f)} = U C^{(e)} U^* \tag{41}$$

$$A^{(f)} = U A^{(e)} U^T \tag{42}$$

Proof. One has:

$$\sum_{i,j} C_{ij}^{(e)} a_{e_i}^+ a_{e_j} = \sum_{i,j} C_{ij}^{(e)} a_{U f_i}^+ a_{U f_j}$$

$$= \sum_{i,j} C_{ij}^{(e)} a_{\sum_h (U)_{hi} f_h}^+ a_{\sum_k (U)_{kj} f_k}$$

$$= \sum_{i,j} C_{ij}^{(e)} \sum_{h,k} (U)_{hi} \overline{(U)_{kj}} a_{f_h}^+ a_{f_k}$$

$$= \sum_{h,k} \sum_{i,j} (U)_{hi} C_{ij}^{(e)} ((U)^*)_{jk} a_{f_h}^+ a_{f_k}$$

$$= \sum_{h,k} (U C^{(e)} (U)^*)_{hk} a_{f_h}^+ a_{f_k}$$

which is equivalent to (41). Similarly, with sum over repeated indexes,

$$A_{ij}^{(e)} a_{e_i}^+ a_{e_j}^+ = A_{ij}^{(e)} a_{U f_i}^+ a_{U f_j}^+ = A_{ij}^{(e)} a_{U_{hi} f_h}^+ a_{U_{kj} f_k}^+$$

$$= A_{ij}^{(e)} a_{U_{hi} f_h}^+ a_{U_{kj} f_k}^+ = U_{hi} A_{ij}^{(e)} U_{kj} a_{f_h}^+ a_{f_k}^+$$

$$= U_{hi} A_{ij}^{(e)} (U^T)_{jk} a_{f_h}^+ a_{f_k}^+ = (U A^{(e)} U^T)_{hk} a_{f_h}^+ a_{f_k}^+$$

which is equivalent to (42).

Definition 2.3. A homogeneous quadratic boson field $X \equiv (A^{(e)}, C^{(e)})$ is called **diagonalizable** if there exists $U \in GL(\mathcal{T})$ and two diagonal matrices D_A, D_C with real entries, such that

$$(U A^{(e)} U^T, U C^{(e)} U^*) = (D_A, D_C)$$

3. Non homogeneous quadratic fields

We know that any non homogeneous quadratic field can be written, up to strict equivalence, in the form

$$F + a_h^+ + a_h + c \tag{43}$$

where F is an homogeneous quadratic field in standard form, i.e. in some fixed linear basis (e_j) of $\mathcal{T}$ and for some complex $d \times d$ matrices $A = (A_{jh})$, $C = (C_{jh})$,

$$F = \sum_{j,h} \left(A_{jh} a_h^+ a_j^+ + \bar{A}_{jh} a_j a_h + C_{jh} a_j^+ a_h \right) \; ; \tag{44}$$

$$A_{jh} = A_{hj} \; , \; \bar{C}_{jh} = C_{hj} \tag{45}$$

3.1. *Equivalence of quadratic fields under automorphisms of $\mathcal{P}_\Gamma(a^+, a, \mathcal{T})$*

Recall that $\mathcal{P}_\Gamma(a^+, a, \mathcal{T})$ is the polynomial algebra generated by the $a_f^\pm$ with $f \in \mathcal{T}$ and denote $\mathrm{Aut}\,(\mathcal{P}_\Gamma(a^+, a, \mathcal{T}))$ its group of $*$–automorphisms.

Definition 3.1. Let G be a group and let

$$w : G \to \mathrm{Aut}\,\big(\mathcal{P}_\Gamma(a^+, a, \mathcal{T})\big)$$

be a representation of G into $\mathrm{Aut}\,(\mathcal{P}_\Gamma(a^+, a, \mathcal{T}))$.
Two quadratic fields F, F'' are called $(G, w))$–**equivalent** if there exists $x \in G$ such that $u_x(F)$ and F' are equivalent, i.e. $u_x(F) = F' + c$ for some $c \in \mathbb{R}$.

3.2. *Action of the translation group on $\mathcal{P}(a^+, a, \mathcal{T})$*

For $x \in \mathbb{C}$, denote $W(x) := e^{\bar{x}a - x a^+}$ the Weyl operator acting on $\Gamma(\mathbb{C})$ and recall the identities

$$W(x)^* a W(x) = a + x \quad ; \quad W(x)^* a^+ W(x) = a^+ + \bar{x} \tag{46}$$

If (e_j) is an ONB of $\mathcal{T} \equiv \mathbb{C}^d$ then the product $\prod_{j \in D} W(x_j e_j)$ is well defined for any $x = (x_j) \in \mathbb{C}^d$ and the map

$$w_x(X) := \prod_{j=1}^{N} W(x_j e_j)^* X \prod_{j=1}^{N} W(x_j e_j) \qquad ; \qquad \forall x \equiv (x_j) \qquad (47)$$

is a $*$–automorphism of $\mathcal{P}_\Gamma(a^+, a, \mathcal{T})$ characterized by the property:

$$w_x(a_{e_j}^+) = a_{e_j}^+ + x_j \qquad ; \qquad w_x(a_{e_j}) = a_{e_j} + \bar{x}_j \qquad ; \qquad \forall j$$

The map

$$x \in \mathcal{T} \mapsto w_x \in \text{Aut}\left(\mathcal{P}_\Gamma(a^+, a, \mathcal{T})\right)$$

is a representation of the additive group $\mathcal{T}$ and in the following $(\mathcal{T}, w)$–equivalence will be simply called **translation equivalence**. Equivalently:

Definition 3.2. Two quadratic fields F, F' are called **translation equivalent** if there exists $x \in \mathcal{T}$ such that $w_x(F)$ and F' are equivalent, i.e. $w_x(F) = F' + c$ for some $c \in \mathbb{R}$.

3.3. Translation equivalence between homogeneous and non–homogeneous quadratic boson fields

In the present section we study **under which conditions a quadratic field is translation equivalent to an homogeneous quadratic field.** Theorem 3.1 below shows that not always a non–homogeneous quadratic field F, of the form given by the right hand side of (49), is translation equivalent to an homogeneous quadratic field. A necessary and sufficient condition for this is the existence of solutions of equation (51) below where A and C are the matrices defining the standard form of the homogeneous quadratic part of F in a linear basis (e_j) of $\mathcal{T}$.

Theorem 3.1. *Let*

$$F = \sum_{i,j} \left(A_{ij} a_{e_i}^+ a_{e_j}^+ + \bar{A}_{ji} a_{e_j} a_{e_i} + C_{ij} a_{e_i}^+ a_{e_j} \right) \qquad (48)$$

be the expression of an homogeneous quadratic field in the linear basis (e_j) of $\mathcal{T}$. Then for any $x \equiv (x_j) \in \mathbb{C}^d$ the translated field $w_{\bar{x}}(F)$ has the same homogeneous quadratic part as F.
If the right hand side of (48) is the standard form of F in the (e_j)–basis, then

$$w_{\bar{x}}(F) = F + a_h^+ + a_h + c \qquad (49)$$

with the constant c given by

$$c := \sum_{j,h} \left(\bar{A}_{jh}\bar{x}_j\bar{x}_h + A_{jh}x_h x_j + C_{jh}x_j\bar{x}_h \right) \in \mathbb{R} \qquad (50)$$

and h satisfying the equation

$$(2A + CJ_e)x = h \qquad (51)$$

where $J_e : \mathcal{T} \to \mathcal{T}$ denotes the anti–linear involution defined by the basis (e_j). i.e.

$$g := \sum_j y_j e_j \;\Rightarrow\; J_e g := \sum_j \bar{y}_j e_j \qquad (52)$$

Conversely, given $h \in \mathcal{T}$, if x is any solution of equation (51), the associated translation automorphism $w_{\bar{x}}$ satisfies (49) with c given by (50).

Proof. In the above notations, under the automorphism $w_{\bar{x}}$, one has

$$a_j \to a_j + \bar{x}_j \qquad ; \qquad a_j^+ \to a_j^+ + x_j$$

Therefore

$$w_{\bar{x}}(F) = \sum_{j,h} \big(A_{jh}(a_h^+ + x_h)(a_j^+ + x_j) + \bar{A}_{jh}(a_j + \bar{x}_j)(a_h + \bar{x}_h)$$

$$+ \, C_{jh}(a_j^+ + x_j)(a_h + \bar{x}_h) \big)$$

$$= \sum_{j,h} \big(A_{jh} a_h^+ a_j^+ + A_{jh} a_h^+ x_j + A_{jh} x_h a_j^+ + A_{jh} x_h x_j \big)$$

$$+ \sum_{j,h} \big(\bar{A}_{jh} a_j a_h + \bar{A}_{jh} a_j \bar{x}_h + \bar{A}_{jh} \bar{x}_j a_h + \bar{A}_{jh} \bar{x}_j \bar{x}_h \big)$$

$$\sum_{j,h} \big(C_{jh} a_j^+ a_h + C_{jh} a_j^+ \bar{x}_h + C_{jh} x_j a_h + C_{jh} x_j \bar{x}_h \big)$$

$$= \sum_{j,h} \big(A_{jh} a_h^+ a_j^+ + \bar{A}_{jh} a_j a_h + C_{jh} a_j^+ a_h \big)$$

$$+ \sum_{j,h} \big(A_{jh} a_h^+ x_j + A_{jh} x_h a_j^+ + C_{jh} a_j^+ \bar{x}_h \big)$$

$$+ \sum_{j,h} \big(\bar{A}_{jh} a_j \bar{x}_h + \bar{A}_{jh} \bar{x}_j a_h + C_{jh} x_j a_h \big)$$

$$+ \sum_{j,h} \big(\bar{A}_{jh} \bar{x}_j \bar{x}_h + A_{jh} x_h x_j + C_{jh} x_j \bar{x}_h \big)$$

Therefore, with c given by (50), one can write

$$w_{\bar{x}}(F) = F + \sum_{j,h} A_{jh} x_j a_{e_h}^+ + \sum_{j,h} A_{jh} x_h a_{e_j}^+ + \sum_{j,h} C_{jh} \bar{x}_h a_{e_j}^+$$

$$+ \sum_{j,h} \bar{A}_{jh} \bar{x}_h a_{e_j} + \sum_{j,h} \bar{A}_{jh} \bar{x}_j a_{e_h} + \sum_{j,h} C_{jh} x_j a_{e_h} + c$$

The 1–st order term can be written in the form

$$\sum_{j,h} A_{jh} x_j a_{e_h}^+ + \sum_{j,h} A_{jh} x_h a_{e_j}^+ + \sum_{j,h} \bar{A}_{jh} \bar{x}_h a_{e_j}$$

$$+ \sum_{j,h} \bar{A}_{jh} \bar{x}_j a_{e_h}$$

$$+ \sum_{j,h} C_{jh} \bar{x}_h a_{e_j}^+ + \sum_{j,h} C_{jh} x_j a_{e_h}$$

and, using the symmetry of A and the Hermiteanity of C this is equal to

$$\sum_{h,j} A_{hj} x_j a_{e_h}^+ + \sum_{j,h} A_{hj} x_j a_{e_h}^+ + \sum_{j,h} \bar{A}_{hj} \bar{x}_j a_{e_h}$$
$$+ \sum_{j,h} \bar{A}_{hj} \bar{x}_j a_{e_h}$$

$$+ \sum_{j,h} C_{hj} \bar{x}_j a_{e_h}^+ + \sum_{j,h} C_{jh} x_j a_{e_h}$$

$$= \sum_{h,j} 2 A_{hj} x_j a_{e_h}^+ + \sum_{j,h} 2 \bar{A}_{hj} \bar{x}_j a_{e_h} + \sum_{j,h} C_{hj} \bar{x}_j a_{e_h}^+$$
$$+ \sum_{j,h} \bar{C}_{hj} x_j a_{e_h}$$

$$= a_{\sum_{h,j} 2 A_{hj} x_j e_h}^+ + a_{\sum_{j,h} 2 A_{hj} x_j e_h} + a_{\sum_{j,h} C_{hj} \bar{x}_j e_h}^+$$
$$+ a_{\sum_{j,h} C_{hj} \bar{x}_j e_h}$$

$$= a_{\sum_{h,j} 2 A_{hj} x_j e_h + \sum_{j,h} C_{hj} \bar{x}_j e_h}^+ + a_{\sum_{j,h} 2 A_{hj} x_j e_h + \sum_{j,h} C_{hj} \bar{x}_j e_h} \qquad (53)$$

Now recall that, in the notation (20) and with J_e being the involution (52),

$$C J_e x = C \bar{x} = \sum_j \bar{x}_j C e_h = \sum_{j,h} C_{hj} \bar{x}_j e_h \qquad ;$$

$$x : = \sum_j x_j e_j$$

and similarly for A. Therefore (53) can be written in the form

$$a_{2Ax + C J_e x}^+ + a_{2Ax + C J_e x} = a_{(2A + C J_e)x}^+ + a_{(2A + C J_e)x}$$

Defining

$$h := (2A + C J_e)x \qquad ; \qquad (54)$$
$$x := \sum_j x_j e_j \quad , \quad h, x \in \mathcal{T} \qquad (55)$$

by construction h is in the range of $(2A + C J_e)$ and one has

$$w_{\bar{x}}(F) = F + a_h^+ + a_h^+ + c$$

which is (49). Conversely, if $h \in Range(2A + C J_e)$ then the first part of the proof shows that, if x is any solution of equation (54), the associated translation automorphism $w_{\bar{x}}$ satisfies (49) and this proves the statement.

Remark. Theorem 3.1 implies that, if the operator $2A + CJ_e$ is surjective (hence, for finite dimensional $\mathcal{T}$, invertible), any quadratic field is equivalent to an homogeneous quadratic field.
If the operator $2A + CJ_e$ is not surjective, only those vectors which are in the range of $2A + CJ_e$ can appear in any translation of the standard form of an homogeneous quadratic field in the (e_j)–basis.

3.4. *C–diagonal standard form of an homogeneous quadratic field*

Lemma 3.1. *Any homogeneous quadratic field can be written in the form*

$$F = \sum_{i,j} \left(A_{ij}^{(f)} a_{f_i}^+ a_{f_j}^+ + \bar{A}_{ji}^{(f)} a_{f_j} a_{f_i} \right) + \sum_i D_{C,i}^{(f)} a_{f_i}^{+2} \tag{56}$$

where (f_i) is an ortho–normal basis of $\mathcal{T}$, the $D_{C,i}$ are real numbers and $A^{(f)}$ is a symmetric matrix. If

$$F = \sum_{i,j} \left(A_{ij}^{(e)} a_{e_i}^+ a_{e_j}^+ + \bar{A}_{ji}^{(e)} a_{e_j} a_{e_i} + C_{ij}^{(e)} a_{e_i}^+ a_{e_j} \right) \tag{57}$$

is the canonical form with respect to an orthonormal basis (e_j), the $D_{C,i}$ are the eigen–values of $C^{(e)}$ and

$$A_{hk}^{(f)} := \begin{cases} (V^* A^{(e)} V^*)_{kk} \,, & \text{if } h = k \\ \frac{1}{2}\left((V^* A^{(e)} V^*)_{hk} + (V^* A^{(e)} V^*)_{kh} \right) \,, & \text{if } h \neq k \end{cases} \tag{58}$$

where V is any unitary that diagonalizes $C^{(e)}$.

Proof. In the standard form of an homogeneous quadratic field with respect to a linear basis (e_j) the choice of this basis is arbitrary. Therefore one can always choose (e_j) to be an orthonormal basis. If V is any unitary, we know from Lemma 2.6 that in the orthonormal basis

$$f_k := \sum_j V_{jk} e_j = V e_k \quad ; \quad V := (V_{jk})$$

the canonical form of F is given by

$$F = \sum_{i,j} \left(A_{ij}^{(f)} a_{f_i}^+ a_{f_j}^+ + \bar{A}_{ji}^{(f)} a_{f_j} a_{f_i} + C_{ij}^{(f)} a_{f_i}^+ a_{f_j} \right) \tag{59}$$

with

$$C^{(f)} = V^* C^{(e)} V \tag{60}$$

and $A^{(f)}$ given by (58). By the spectral theorem one can choose the unitary V so that $V^* C^{(e)} V$ is diagonal. With this choice the identity (59) can be written in the form (56).

3.5. A–diagonal canonical form of an homogeneous quadratic field

A different canonical form of an homogeneous quadratic field can be obtained exploiting the Autonne–Takagi decomposition [5], [10], of a symmetric matrix, an idea originally due to A. Chebotarev and A. Teretenkov [7].

Theorem 3.2. *Let A be a symmetric matrix, then there exists a unitary matrix V and a real non–negative diagonal matrix D such that*

$$A = V D_A V^T \quad ; \quad D_A \geq 0 \ diagonal \tag{61}$$

where the columns of V are an orthonormal set of eigenvectors for AA^* and the corresponding diagonal entries of D_A are the non–negative square roots of the corresponding eigenvalues of A^*A. is called the Autonne–Takagi decomposition of A.

Proof. See the book [9].

Lemma 3.2. *Any homogeneous quadratic field can be written in the form*

$$F = \sum_i D_{A,i} \left(a^+_{V f_i} a^+_{V f_i} + a^-_{V f_i} a^-_{V f_i} \right) + \sum_i D_i a^{+2}_{f_i} \tag{62}$$

where (f_i) is an ortho–normal basis of $\mathcal{T}$, the $D_{A,i}$ are positive real numbers, the $D_{C,i}$ are real numbers and V is a unitary operator. If

$$F = \sum_{i,j} \left(A_{ij} a^+_{f_i} a^+_{f_j} + \bar{A}^{(f)}_{ji} a_{f_j} a_{f_i} \right) + \sum_i D_{C,i} a^{+2}_{f_i} \tag{63}$$

is the C–diagonal canonical form of F with respect to an orthonormal basis (e_j), the $D_{A,i}$ are the diagonal elements of D_A and V is the unitary operator in the Autonne–Takagi decomposition (61) of A.

Proof. From Lemma 3.1 it is known that any homogeneous quadratic field can be written in the form

$$F = \sum_{i,j} \left(A_{ij} a^+_{f_i} a^+_{f_j} + \bar{A}_{ji} a_{f_j} a_{f_i} \right) + \sum_i D_i a^{+2}_{f_i}$$

where (f_i) is an ortho–normal basis of $\mathcal{T}$, the $D_{C,i}$ are real numbers and $A \equiv (A_{ij})$ is a symmetric matrix. If V is the unitary operator appearing in the Takagi decomposition (61) of A, one has

$$A_{ij}a^+_{f_i}a^+_{f_j} = (VD_AV^T)_{ij}a^+_{f_i}a^+_{f_j} = V_{ih}D_{A,h}(V^T)_{hj}a^+_{f_i}a^+_{f_j}$$
$$= D_{A,h}a^+_{V_{ih}f_i}a^+_{(V^T)_{hj}f_j}$$

$$= D_{A,h}a^+_{V_{ih}f_i}a^+_{V_{jh}f_j} = D_{A,h}a^+_{Vf_h}a^+_{Vf_h}$$

and this implies (62).

3.6.   The Benedetti–Cragnolini canonical form

We have seen that homogeneous quadratic boson fields on $\mathcal{T} \equiv \mathbb{C}^d$ are parametrized by pairs (A, C), where A is a symmetric and C an Hermitean $d \times d$ matrix, and that on these pairs the group $GL(\mathcal{T})$ acts through the transformations (41), (42), i.e.

$$(A, C) \mapsto (UAU^T, UCU^*) \quad ; \quad U \in GL(\mathcal{T}) \tag{64}$$

The paper [6] considers a very similar action on the same space of pairs, namely

$$(A, C) \mapsto (U^TAU, U^*CU) \quad ; \quad U \in GL(\mathcal{T}) \tag{65}$$

and, with respect to this action, they prove the following theorem.

Theorem 3.3. *If $A, C \in M_d(\mathbb{C})$, $A = A^T$, and $C^* = C$, C strictly positive definite, the orbit of the pair (A, C) under the action (65) contains a point of the form $(D, 1)$ where D is a diagonal matrix with nonnegative entries.*

Remark. Denoting

$$V := U^T$$

the action (65) becomes

$$(A, C) \mapsto (VAV^T, (V^T)^*CV^T) = (VAV^T, \overline{V}CV^T)$$

Recall that conjugation

$$j : M \in M_d(\mathbb{C}) \to j(M) := \overline{M}$$

is an anti–linear algebra automorphism of $M_d(\mathbb{C})$ commuting both with the transpose and conjugate operations and leaving real matrices fixed and $GL(\mathcal{T})$ invariant. Therefore, defining the action

$$(id, j) : (A, C) \mapsto (A, j(C)) \tag{66}$$

the action resulting from left composition of (id, j) with the action (65) has the form

$$(A, C) \;\mapsto\; (VAV^T, j(\overline{V}CV^T)) = (VAV^T, Vj(C)j(V^T))$$

$$= (VAV^T, V\overline{C}V^*) \tag{67}$$

and its orbits will still enjoy the property described in Theorem 3.3.
In conclusion: the Benedetti–Cragnolini action (65), up to left composition with the extended conjugation (id, j) given by (66), differs from the quadratic boson field action (64) only by the replacement of C by $\overline{C}$ in the final result.
Furthermore notice that, if $C = C^*$, then

$$(\overline{C})^* = \overline{(C^T)} = \overline{(C^*)} = \overline{C}$$

i.e. conjugation maps Hermitean elements of $M_d(\mathbb{C})$ in themselves.

Corollary 3.1. *If $A, C \in M_d(\mathbb{C})$, $A = A^T$, and $C^* = C$, C strictly positive definite **and with real entries**, the orbit of the pair (A, C) under the action (64) contains a point of the form $(D, 1)$ where D is a diagonal matrix with nonnegative entries.*

Proof. Since the actions (64) and (67) coincide on matrices with real entries, the thesis follows from the above remark. □

Remark. Notice that if $U \in GL(\mathcal{T})$ is such that

$$(U^T A U, U^* C U) = (D, 1) \tag{68}$$

then denoting $C - V_C^* D_C V_C$ the spectral decomposition of C, one has

$$U^* C U = 1 \iff C = (U^*)^{-1} U^{-1} \iff V_C^* D_C V_C - (U^*)^{-1} U^{-1}$$

$$\iff 1 = D_C^{1/2} V_C (U^{-1})^* U^{-1} V_C^* D_C^{1/2}$$

where $D_C^{1/2}$ is the positive square root of D_C. Hence $U^{-1} V_C^* D_C^{1/2}$ is an isometry so, in finite dimensions, a unitary. In this case $D_C^{1/2} V_C U$ is a unitary.
In this sense one can say that U is a *quasi–unitary*.

References

1. Luigi Accardi, Andreas Boukas: Canonical forms of quadratic Boson fields, QBIC 2018
2. Luigi Accardi, Andreas Boukas, Yun-Gang Lu: The vacuum distributions of the truncated Virasoro fields are products of Gamma distributions, OSID (2017)
3. Luigi Accardi, Andreas Boukas: Fourier transform of random variables associated with the multi-dimensional Heisenberg Lie algebra, Proceedings of the American Mathematical Society, Proc. Amer. Math. Soc. 143 (2015) 4095-4101
4. L. Accardi, I. Ya. Areféva , I. V. Volovich: Fermionic Meixner classes, Lie algebras and quadratic Hamiltonians Indian J. Pure Appl. Math. 46 (4) (2015) 517-538
5. Autonne, L.: Sur les matrices hypo–hermitiennes et sur les matrices unitaires, Ann. Univ. Lyon, 38 (1915) 1–77
6. Benedetti, R.; Cragnolini, P.: On simultaneous diagonalization of one Hermitian and one symmetric form, Linear Algebra Appl. 57 (1984) 215–226,
7. Alexander M. Chebotarev, Alexander E. Teretenkov: Singular value decomposition for the Takagi factorization of symmetric matrices Applied Mathematics and Computation 234 (2014) 380–384
8. Feinsilver-Pap2007 P. J. Feinsilver and G. Pap, Calculation of Fourier transforms of a Brownian motion on the Heisenberg group using splitting formulas, Journal of Functional Analysis 249 (2007) 30.
9. Horn, Roger A.; Johnson, Charles R.: Matrix analysis (2nd ed.), Cambridge University Press (2013) ISBN 978-0-521-54823-6
10. Takagi, T.: On an algebraic problem related to an analytic theorem of Carathédory and Fejér and on an allied theorem of Landau, Japan. J. Math. 1 (1925) 83–93

Quantum Bio-Informatics VI
© 2020 World Scientific Publishing Co. Pte. Ltd.
pp. 29–36

ALGORITHMS FOR BROWNIAN DYNAMICS SIMULATION

TADASHI ANDO

RIKEN QBiC, International Medical Device Alliance (IMDA)
6F, 1-6-5 Minatojima-minamimachi, Chuo-ku, Kobe, Hyogo 650-0047, Japan
E-mail: tadashi.ando@riken.jp

YUJI SUGITA

RIKEN QBiC, International Medical Device Alliance (IMDA)
6F, 1-6-5 Minatojima-minamimachi, Chuo-ku, Kobe, Hyogo 650-0047, Japan
RIKEN Advanced Institute for Computational Science,
7-1-26 Minatojima-minamimachi, Chuo-ku, Kobe, Hyogo 650-0047, Japan
RIKEN Theoretical Molecular Science Laboratory and iTHES,
2-1 Hirosawa, Wako-shi, Saitama 351-0198, Japan
E-mail: sugita@riken.jp

Brownian dynamics (BD) is one of the most important computational techniques for simulating the motion of macromolecules in a fluid environment. Since an integration scheme for BD was firstly developed by Ermak and McCammon in 1978 [D. L. Ermak, J. A. McCammon, J Chem Phys 69, 1352 (1978)], several different integration algorithms have been introduced. In this work, we evaluated algorithmic efficiency of BD algorithms by simulating simple dimer particles connected with a harmonic spring for various time steps in the absence of hydrodynamic interactions. Within the examined four schemes, a BD algorithm involving a colored noise gave the best efficiency in terms of computational cost without significant loss of numerical accuracy. It was 20 times more efficient than the original BD algorithm.

1. Introduction

Brownian dynamics (BD) is one of the most important computational techniques for simulating the motion of macromolecules in a fluid environment [1,2]. BD has been used with coarse-grained models, where solvent molecules are not treated explicitly; rather, their dynamical effects on a solute molecule are incorporated in a stochastic manner consistent with hydrodynamics. Now, BD is widely used for simulating long-time dynamics of large systems that all-atom molecular dynamics (MD) including water molecules

cannot feasibly access, as well as for extracting essential features of macro-
molecular dynamics of complex systems. For examples, intracellular environments represented by coarse-grained macromolecules were recently
simulated by BD to investigate dynamics of macromolecules in vivo [3,4].

An integration scheme for BD was firstly developed by Ermak and Mc-
Cammon in 1978 [5]. The algorithm is a first-order algorithm for approximating numerical solution of stochastic differential equations, which is called
Euler-Maruyama method in mathematics [6]. Since the landmark work, several new ideas have been introduced to improve the performance of the
original BD algorithm. van Gunsteren and Berendsen proposed an algorithm where forces on particles are described as a power series to correct
drift of forces during a finite time step [7]. Iniesta and García de la Torre
developed a second-order stochastic Runge-Kutta algorithm for BD simulations [8]. The algorithm requires more than one calculation of energy
and force of the particles per time step, which increase computational cost
per step. But, this is compensated by being able to employ a longer time
step. Recently, Leimkuhler and Matthews introduced a simple modification of the original BD algorithm involving a colored noise. The key idea
was arrived by introducing the large friction limit in a particular numerical
method for Langevin dynamics [9,10].

In the present work, we evaluated algorithmic efficiency of BD algorithms mentioned above by simulating simple dimer particles. Details of
the BD algorithms and a model used for the evaluation are described in
section 2. The results and discussion are presented in section 3, and conclusions are presented in section 4.

2. Model and Methods

2.1. *Brownian Dynamics Algorithms*

Derivation of Brownian dynamics algorithm starts from the Langevin equation given by [5]

$$m_i \frac{d^2 r_i}{dt^2} = -\zeta_i \frac{dr_i}{dt} + F_i + R_i. \tag{1}$$

Here, m_i and $\mathbf{r}_i$ represent the mass and position of particle i, respectively.
ζ_i is a frictional coefficient and is determined by the Stokes' law, that is, ζ_i
$= 6\pi a_i \eta$ in which a_i is a Stokes radius of particle i and η is the viscosity
of water. $\mathbf{F}_i$ is the systematic force on particle i. $\mathbf{R}_i$ is a random force on
particle i having a zero mean $<\mathbf{R}_i(t)> = 0$ and a variance $<\mathbf{R}_i(t)\mathbf{R}_j(0)>$
$= 2\zeta_i k_B T \delta_{ij} \delta(t) \mathbf{I}$ where k_B is the Boltzmann's constant, T is absolute

temperature, δ_{ij} is the Kronecker delta, $\delta(t)$ is the Dirac delta, and $\mathbf{I}$ is 3 × 3 unit tensor; this derives from the dissipation-fluctuation theorem.

For the overdamped limit, that is, the solvent damping is large and the inertial memory is lost in a very short time, we set the left side of Eq. 1 to zero,

$$\zeta_i \frac{dr_i}{dt} = F_i + R_i. \tag{2}$$

A first-order approximation for the solution of Eq. 2 is called Brownian dynamics;

$$r_i(t + \Delta t) = r_i(t) + \frac{\Delta t}{\zeta_i} F_i + \sqrt{\frac{2k_B T}{\zeta_i} \Delta t} z_i, \tag{3}$$

where Δt is a time step and $\mathbf{z}_i$ is a random noise vector obtained from Gaussian distribution. This BD algorithm was introduced by Ermak and McCammon in 1978 [5], which is the most common technique. Hereafter, we call Eq. 3 "EM" algorithm.

An extension of EM algorithm is to permit force $\mathbf{F}_i$ to vary linearly over the time step Δt,

$$r_i(t + \Delta t) = r_i(t) + \frac{\Delta t}{\zeta_i} \left(F_i + \frac{1}{2}\dot{F}_i \Delta t \right) + \sqrt{\frac{2k_B T}{\zeta_i} \Delta t} z_i. \tag{4}$$

Here, $\dot{F}_i$ is the derivative of the force at the time t and is numerically approximated by

$$\dot{F}_i(t) = [F_i(t) - F_i(t - \Delta t)]/\Delta t. \tag{5}$$

This algorithm was introduced by van Gunsteren and Berendsen in 1982 [7]. Hereafter, we call Eqs. 4 and 5 "GB" algorithm.

EM algorithm corresponds to an explicit Euler or first-order Runge-Kutta integration scheme. Iniesta and García de la Torre developed a second order stochastic Runge-Kutta algorithm for BD simulations (IG) [8]

$$r_i(t + \Delta t) = r_i(t) + \frac{\Delta t}{2\zeta_i} \left(F_i^1 + F_i^2 \right) + \sqrt{\frac{2k_B T}{\zeta_i} \Delta t} z_i, \tag{6}$$

where

$$F_i^1 = F_i[r(t)], \tag{7}$$

and

$$F_i^2 = F_i[r'(t)]. \tag{8}$$

32

Here,

$$r_i'(t) = r_i(t) + \frac{\Delta t}{2\zeta_i} F_i^1 + \sqrt{\frac{2k_BT}{\zeta_i}} \Delta t z_i. \tag{9}$$

Recently, Leimkuhler and Matthews introduced a simple modification of the EM method for BD involving a colored noise [9,10]. The scheme described by

$$r_i(t + \Delta t) = r_i(t) + \frac{\Delta t}{\zeta_i} F_i + \sqrt{\frac{k_BT}{2\zeta_i}} \Delta t \left(z_i(t) + z_i(t + \Delta t) \right). \tag{10}$$

Since the method use the colored noise though the noise decorrelates in just a couple of time steps, this is no longer a Markov process. They originally called the scheme "BAOAB" limit method. Here, to unify the names of algorithms, we call Eq. 10 "LM" method.

2.2. Simulation Model and Analysis

In order to compare the efficiency of the above BD algorithms, we performed BD simulations of a model molecule consisting of two beads with radii of a connected with a harmonic spring. The spring energy of the model is defined by

$$U = \frac{1}{2} \frac{k_BT}{(x_0\delta)^2} (x - x_0)^2. \tag{11}$$

Here, x and x_0 are the instantaneous and equilibrium lengths between two particles, respectively, and δ is the stiffness parameter. With this formulation, the variance of the length x is known as [11]

$$\frac{\left\langle (x - \langle x \rangle)^2 \right\rangle}{x_0^2} = \frac{\delta^2 + 3\delta^6}{\left(1 + \delta^2\right)^2} \approx \delta^2, \tag{12}$$

where $<...>$ represents averaging over the simulation time. In the calculations, distance, time, and energy were non-dimensionalized by a, $\tau_B = a^2/D$, and k_BT, respectively, where D is the diffusion coefficient of a particle calculated by Stoke-Einstein relationship, $D = k_BT/6\pi a\eta$. x_0 of 2 and $\delta = 0.1$ were used. Time steps of 5×10^{-5}, 1×10^{-4}, 1×10^{-3}, 2×10^{-3}, 4×10^{-3}, 6×10^{-3}, 8×10^{-3}, 1×10^{-2}, and 2×10^{-2} were considered. BD simulations were performed for 10^4 in simulation time with each condition.

For evaluating numerical accuracy of the algorithms in various time steps, we analyzed three quantities:

$$\text{Error}\left[V\left(x\right)\right] = \frac{V\left(x\right) - V\left(x\right)^{*}}{V\left(x\right)^{*}}, \tag{13}$$

$$\text{Error}\left[\langle U \rangle\right] = \frac{\langle U \rangle - \langle U \rangle^{*}}{\langle U \rangle^{*}}, \tag{14}$$

$$\text{Error}\left[V\left(U\right)\right] = \frac{V\left(U\right) - V\left(U\right)^{*}}{V\left(U\right)^{*}}. \tag{15}$$

Here, $V(x)$ and $V(U)$ represent variance of length x and energy U, respectively. For $V(x)^{*}$, the theoretical value given by Eq. 12 was used. For $<U>^{*}$ and $V(U)^{*}$, the values obtained from the BD simulation with Δt of 5×10^{-5} in the EM algorithm were used. These three quantities represent relative errors from an ideal value or values obtained from the simulation with a sufficiently small time step.

3. Results and Discussion

In Fig. 1, the relative errors in $V(x)$, $<U>$, and $V(U)$ are shown. For the EM algorithm, the errors continuously increased with increase of time step. The GB algorithm slightly improved in numerical accuracy, though the errors suddenly raised when $\Delta t > 0.01$ were used in our model. Although the errors in the IG algorithm was slightly smaller than those in the GB algorithm, the large errors were still observed for the large time steps. For the LM algorithm, in contrast to the EM, GB, and IG algorithms, the relative errors were close to zero over the time steps examined in this work. We tried to use Δt of 0.04 with the LM algorithm, but it was very unstable and the errors were too large. In our model, if we kept the relative errors less than 0.1, the maximum time steps for the algorithms of EM, GB, IG, and LM are 0.001, 0.005, 0.008, and 0.02, respectively.

We also checked diffusion coefficient of the model dimer to evaluate numerical accuracy of the algorithms. Although the significant increase of errors in energy and bond distance for large time steps were observed in the EM, GB and IG algorithms as shown in Fig. 1, we could not see any clear trend in diffusion coefficients of the molecule with time step.

We discuss the results from the point of view of computational efficiency.

Generally speaking, energy/force calculation is the most time consuming part in conventional molecular simulations, e.g. molecular dynamics and BD without hydrodynamic interactions.

The EM, GB, and LM methods perform one energy/force calculation per step. On the other hand, the IG algorithm requires two energy/force calculations per step. Therefore, although the IG algorithm can adopt a time step about two times larger than that of the GB algorithm, computational costs of these algorithms would be almost the same.

As a consequence, we could say that the algorithmic efficiency of the examined BD schemes, EM, GB, IG, and LM, are roughly in the ratio of 1:5:4:20.

In this work, hydrodynamic interactions (HI) between particles were ignored. HI are interactions between the particles via solvent flow, which give rise to correlated motions between particles [1,2]. Therefore, in BD simulations with HI, long time steps would be employed compared to BD simulations without HI.

We also used only a small dimer molecule as a model for the evaluation. From a practical point of view, use of multi-particle systems or long-chain polymers would be necessary to systematically assessing the numerical accuracy in BD simulations.

Considering these factors in the evaluation is our next study. However, we believe that the same trend in algorithmic efficiency observed in this work would be seen even in the larger systems in the presence of HI.

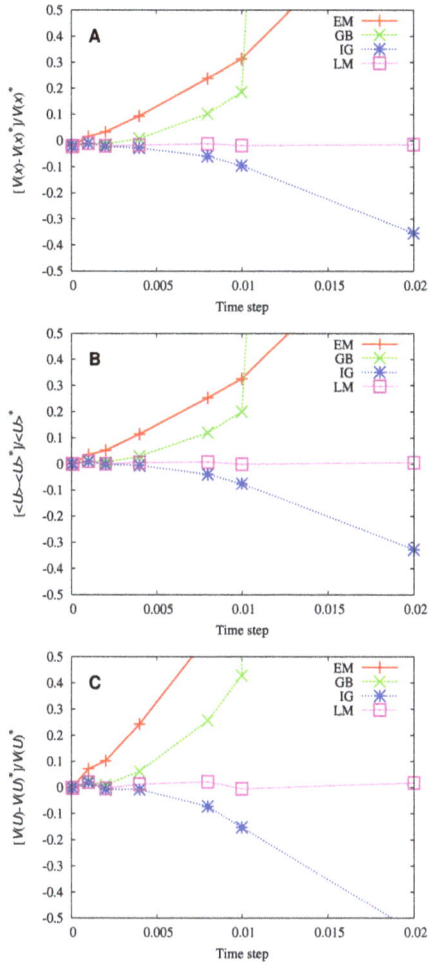

Figure 1. Relative errors in variance of bond distance (A), average energy (B), and variance of energy (C) as a function of time step in various Brownian dynamics simulation algorithms.

4. Conclusions

In this work, we evaluated algorithmic efficiency of four existing BD algorithms by simulating simple dimer particles in the absence of HI. Our result clearly showed that a BD algorithm involving a colored noise introduced by Leimkuhler and Matthews is the most efficient scheme in term of computational cost without significant loss of numerical accuracy.

Acknowledgments

We thank Yasuhiro Matsunaga for his comments on the manuscript.

References

1. M. P. Allen and D. J. Tildesley Computer simulation of liquids (Clarendon Press) (1989).
2. T. Schlick Molecular Modeling and Simulation: An Interdisciplinary Guide: An Interdisciplinary Guide (Springer Science & Business Media) (2010).
3. T. Ando and J. Skolnick, Proc. Natl. Acad. Sci. USA. 107, 18457 (2010).
4. S. R. McGuffee and A. H. Elcock, PLoS Comput. Biol. 6, (2010).
5. D. L. Ermak and J. A. McCammon, J. Chem. Phys. 69, 1352 (1978).
6. P. E. Kloeden and E. Platen Numerical solution of stochastic differential equations (Springer Science & Business Media) (1992).
7. W. F. van Gunsteren and H. J. C. Berendsen, Mol. Phys. 45, 637 (1982).
8. A. Iniesta and J. G. de la Torre, J. Chem. Phys. 92, 2015 (1990).
9. B. Leimkuhler and C. Matthews, J. Chem. Phys. 138, (2013).
10. B. Leimkuhler and C. Matthews, Appl. Math. Res. Express 2013, 34 (2013).
11. K. Klenin, H. Merlitz, and J. Langowski, Biophys. J. 74, 780 (1998).

Quantum Bio-Informatics VI
© 2020 World Scientific Publishing Co. Pte. Ltd.
pp. 37–46

MEMORY KERNELS FOR QUBIT EVOLUTION*

DARIUSZ CHRUŚCIŃSKI

Institute of Physics, Faculty of Physics, Astronomy and Informatics
Nicolaus Copernicus University, Grudziadzka 5/7, 87–100 Toruń, Poland
E-mail: darch@fizyka.umk.pl

We analyze a quantum evolution of the qubit within memory kernel approach.
We provide sufficient conditions which guarantee that the corresponding memory
kernel generates physically legitimate quantum evolution.

Keywords: Open systems; quantum dynamics; non-Markovian evolution.

1. Introduction

The dynamics of open quantum systems attracts nowadays increasing attention [1,2,3] due to the growing interest in controlling quantum systems and modern applications to quantum technologies such as quantum communication, cryptography and computation [4,5]. As is well known [1,2] dynamics of open quantum systems plays important role in the analysis of various phenomena like dissipation, decoherence and dephasing. The usual approach to the dynamics of an open quantum system consists of applying the Born-Markov approximation [1] which leads to the local master equation for the Markovian semigroup

$$\dot{\rho}_t = L[\rho_t] \, , \tag{1}$$

where ρ_t is the density matrix of the investigated system and L is the time-independent generator of the dynamical semigroup defined as follows

$$L[\rho] = -i[H, \rho] + \frac{1}{2} \sum_\alpha \left([V_\alpha, \rho V_\alpha^\dagger] + [V_\alpha \rho, V_\alpha^\dagger] \right) , \tag{2}$$

where H denotes the effective system Hamiltonian, and V_α represent so-called noise operators [6,7]. One calls L defined by (2) the Gorini-

---

*Dedicated to Professor Masanori Ohya.

Kossakowski-Sudarshan-Lindblad form (GKSL) [6,7]. Basically there are two ways to go beyond the standard Markovian master equation (3):

- a time-local master equation

$$\dot{\rho}_t = L_t[\rho_t] \,, \tag{3}$$

  with time dependent local generator L_t,
- and non-local Nakajima-Zwanzig [8] (see also Refs. 9, 10)

$$\dot{\rho}_t = \int_0^t K_{t-\tau}[\rho_\tau]d\tau, \tag{4}$$

  in which quantum memory effects are taken into account through the introduction of the memory kernel K_t. This means that the rate of change of the state ρ_t at time t depends on its history (starting at $t = 0$).

In this paper we analyze the structure of the memory kernel K_t which guarantee that the solution of (4) gives rise to the legitimate dynamical map $\rho \to \rho_t = \Lambda_t[\rho]$, that is, Λ_t is completely positive and trace-preserving (CPTP). This problem was recently analyzed by many authors Refs. 11–23. However, up to now the general solution is not known. In what follows we analyze only the most popular case, namely a qubit evolution.

2. Qubit dephasing

Consider a qubit dephasing described by the following time-local generator

$$L_t[\rho] = \frac{1}{2}\gamma(t)(\sigma_3\rho\sigma_3 - \rho), \tag{5}$$

where $\gamma(t)$ is a real function such that $\Gamma(t) = \int_0^t \gamma(u)du \geq 0$. One finds for the dynamical map $\Lambda_t = \exp(\int_0^t L_u du)$

$$\Lambda_t[\rho] = \frac{1}{2}\left(1 + e^{-\Gamma(t)}\right)\rho + \frac{1}{2}\left(1 - e^{-\Gamma(t)}\right)\sigma_3\rho\sigma_3 \,. \tag{6}$$

The same evolution may be equivalently described by the following memory kernel

$$K_t[\rho] = \frac{1}{2}\kappa(t)(\sigma_3\rho\sigma_3 - \rho), \tag{7}$$

where $\kappa(t)$ is a real function. Now, the question is about the properties of $\kappa(t)$ such that the solution to (4) provides a CPTP map Λ_t. One easily finds the spectrum of K_t:

$$K_t[\sigma_0] = K_t[\sigma_3] = 0 \,, \quad K_t[\sigma_i] = -\kappa(t)\sigma_i \,; \quad i = 1, 2, \tag{8}$$

and hence

$$\Lambda_t[\sigma_0] = \Lambda_t[\sigma_3] = 0 \ , \quad \Lambda_t[\sigma_i] = \lambda(t)\sigma_i \ ; \quad i = 1, 2, \tag{9}$$

where the time-dependent eigenvalue $\lambda(t)$ satisfies

$$\dot{\lambda}(t) = -\int_0^t \kappa(t-\tau)\lambda(\tau)d\tau \ , \quad \lambda(0) = 1 \ . \tag{10}$$

Performing the Laplace transform $\tilde{\lambda}(s) = \int_0^\infty e^{-st}\lambda(t)dt$ one finds

$$\tilde{\lambda}(s) = \frac{1}{s + \tilde{\kappa}(s)} . \tag{11}$$

Note, that the corresponding evolution is given by

$$\rho \ \longrightarrow \ \rho_t = \begin{pmatrix} \rho_{11} & \lambda(t)\rho_{12} \\ \lambda(t)\rho_{21} & \rho_{22} \end{pmatrix}, \tag{12}$$

and hence the map is CPTP iff $|\lambda(t)| \leq 1$.

Example 2.1. Taking $\kappa(t) = \omega^2 H(t)$, where $H(t)$ is a Heaviside function, one obtains

$$\tilde{\lambda}(s) = \frac{s}{s + \omega^2}, \tag{13}$$

and hence $\lambda(t) = \cos(\omega t)$ which obviously satisfies $|\lambda(t)| \leq 1$. Interestingly, the corresponding time-local description gives rise to a highly singular generator with

$$\gamma(t) = -\frac{\dot{\lambda}(t)}{\lambda(t)} = \tan(\omega t) \ . \tag{14}$$

This example shows that the same evolution may be described by a perfectly regular memory kernel with $\kappa(t) = \omega^2 H(t)$ and highly singular local generator with $\gamma(t) = \tan(\omega t)$.

Example 2.2. Consider $\kappa(t) = a^2 e^{-2bt}$, with $b \geq 0$ for $t \geq 0$. One obtains

$$\tilde{\lambda}(s) = \frac{s + 2b}{s(s + 2b) + a^2} = \frac{s + 2b}{(s+b)^2 + c^2}, \tag{15}$$

and we assume $c^2 = a^2 - b^2 > 0$. Inverting into the time domain one finds

$$\lambda(t) = e^{-bt}\left[\cos(ct) + \frac{b}{c}\sin(ct)\right] . \tag{16}$$

It is clear that $|\lambda(t)| \leq 1$ is no longer guaranteed. In particular, if $c = 0$ (i.e. $a = b$) one has

$$\lambda(t) = e^{-at}\left[1 + at\right], \tag{17}$$

which gives rise to the legitimate evolution due to $\lambda(t) \leq 1$. In this case the time-local approach gives

$$\gamma(t) = -\frac{\dot{\lambda}(t)}{\lambda(t)} = \frac{a^2 t}{1 + at} \tag{18}$$

and hence $\Gamma(t) = at - \ln[1 + at]$ implies $e^{-\Gamma(t)} = (1 + at)e^{-at}$.

To find suitable conditions for $\kappa(t)$ consider the following parametrization

$$\lambda(t) = 1 - \int_0^t f(u)du, \tag{19}$$

where $\int_0^t f(u)du \in [0, 2]$. It implies

$$\tilde{\kappa}(s) = -\frac{s\tilde{f}(s)}{1 - \tilde{f}(s)}. \tag{20}$$

The corresponding dynamical map may be represented as follows

$$\Lambda_t[\rho] = \left(1 - \int_0^t f(u)du\right)\rho + \int_0^t f(u)du\ \sigma_3\rho\sigma_3, \tag{21}$$

and hence, if $0 \leq \int_0^t f(u)du \leq 1$ the map Λ_t is a convex combination of identity channel $\mathbb{1}[\rho] = \rho$ and the unitary channel $\mathcal{E}[\rho] = \sigma_3\rho\sigma_3$. In particular if $f(t)$ is a so-called waiting time distribution, i.e $f(t) \geq 0$ and $\int_0^\infty f(t)dt = 1$, then one calls $g(t) = 1 - \int_0^t f(u)du$ a survival probability and gets

$$\Lambda_t[\rho] = g(t)\rho + [1 - g(t)]\ \sigma_3\rho\sigma_3, \tag{22}$$

with $g(t) \in [0, 1]$ and $g(0) = 1$. Such evolution is called *quantum semi-Markov dynamics* [14,18,19]. Note, that taking waiting time distribution $f(t) = \gamma e^{-\gamma t}$ one recovers Markovian semigroup

$$\Lambda_t[\rho] = \frac{1}{2}\left(1 + e^{-\gamma t}\right)\rho + \frac{1}{2}\left(1 - e^{-\gamma t}\right)\sigma_3\rho\sigma_3, \tag{23}$$

with $g(t) = \frac{1}{2}\left(1 + e^{-\gamma t}\right)$.

3. Qubit random unitary evolution

Let us consider a dynamical map Λ_t defined by

$$\Lambda_t[\rho] = \sum_{\alpha=0}^{3} p_\alpha(t)\ \sigma_\alpha\rho\sigma_\alpha, \tag{24}$$

where σ_α are Pauli matrices (with $\sigma_0 = \mathbb{I}$) and $p_\alpha(t)$ stands for a probability distribution. Initial condition $\Lambda_{t=0} = \mathbb{1}$ implies $p_\alpha(0) = \delta_{\alpha 0}$. This evolution is generated by the following time-local generator[24,25]

$$L_t[\rho] = \sum_{k=1}^{3} \gamma_k(t)\left(\sigma_k \rho \sigma_k - \rho\right), \tag{25}$$

with time-dependent decoherence rates $\gamma_k(t)$. Due to the fact that $[L_t, L_u] = 0$ one has $\Lambda_t = \exp(\int_0^t L_\tau d\tau)$ and hence

$$p_\alpha(t) = \frac{1}{4} \sum_{\beta=0}^{3} H_{\alpha\beta} \lambda_\beta(t) , \tag{26}$$

where $H_{\alpha\beta}$ is the Hadamard matrix

$$H = \begin{pmatrix} 1 & 1 & 1 & 1 \\ 1 & 1 & -1 & -1 \\ 1 & -1 & 1 & -1 \\ 1 & -1 & -1 & 1 \end{pmatrix}, \tag{27}$$

and $\lambda_\beta(t)$ are time-dependent eigenvalues of Λ_t

$$\Lambda_t[\sigma_\alpha] = \lambda_\alpha(t)\sigma_\alpha , \tag{28}$$

defined by $\lambda_0(t) = 1$ and

$$\begin{aligned} \lambda_1(t) &= \exp(-2[\Gamma_2(t) + \Gamma_3(t)]), \\ \lambda_2(t) &= \exp(-2[\Gamma_1(t) + \Gamma_3(t)]), \\ \lambda_3(t) &= \exp(-2[\Gamma_1(t) + \Gamma_2(t)]), \end{aligned} \tag{29}$$

with $\Gamma_k(t) = \int_0^t \gamma_k(\tau)d\tau$. Now, the map (24) is CP iff $p_\alpha(t) \geq 0$ which is equivalent to the following set of conditions for λs [24,26]

$$1 + \lambda_1(t) + \lambda_2(t) + \lambda_3(t) \geq 0 , \tag{30}$$

and

$$\begin{aligned} \lambda_1(t) + \lambda_2(t) &\leq 1 + \lambda_3(t), \\ \lambda_3(t) + \lambda_1(t) &\leq 1 + \lambda_2(t), \\ \lambda_2(t) + \lambda_3(t) &\leq 1 + \lambda_1(t). \end{aligned} \tag{31}$$

Now, let as consider a non-local memory kernel defined by

$$K_t[\rho] = \sum_{i=1}^{3} k_i(t)\left(\sigma_i \rho \sigma_i - \rho\right), \tag{32}$$

where $k_i(t)$ ($i = 1, 2, 3$) represent non-trivial memory effects. Note, that the master equation

$$\dot{\Lambda}_t = \int_0^t K_{t-\tau} \Lambda_\tau d\tau \ , \tag{33}$$

considerably simplifies after performing the Laplace transform

$$\tilde{\Lambda}_s = \frac{1}{s - \tilde{K}_s}, \tag{34}$$

where $\tilde{\Lambda}_s := \int_0^\infty e^{-st} \Lambda_t dt$ and similarly for $\tilde{K}_s$. Denoting by $\kappa_\alpha(t)$ the eigenvalues of K_t

$$K_t[\sigma_\alpha] = \kappa_\alpha(t)\sigma_\alpha \ , \tag{35}$$

equation (33) gives rise to the following set of equations

$$\dot{\lambda}_i(t) = \int_0^t \kappa_i(t - \tau)\lambda_i(\tau)\tau, \quad i = 1, 2, 3. \tag{36}$$

Note, that $\kappa_0(t) = 0$ and hence $\lambda_0(t) = 1 = \text{const.}$ In terms of the Laplace transforms $\tilde{\lambda}_i(s)$ and $\tilde{\kappa}_i(s)$ one finds

$$\tilde{\lambda}_i(s) = \frac{1}{s - \tilde{\kappa}_i(s)} \ . \tag{37}$$

In terms of $\tilde{\lambda}_i(s)$ conditions (30)–(31) may be equivalently reformulated as follows: the following functions

$$\ell_0(s) := \frac{1}{s} + \tilde{\lambda}_1(s) + \tilde{\lambda}_1(s) + \tilde{\lambda}_2(s) \ ,$$

$$\ell_1(s) := \frac{1}{s} + \tilde{\lambda}_1(s) - \tilde{\lambda}_3(s) - \tilde{\lambda}_2(s) \ ,$$

$$\ell_2(s) := \frac{1}{s} + \tilde{\lambda}_2(s) - \tilde{\lambda}_1(s) - \tilde{\lambda}_3(s) \ , \tag{38}$$

$$\ell_3(s) := \frac{1}{s} + \tilde{\lambda}_3(s) - \tilde{\lambda}_1(s) - \tilde{\lambda}_2(s) \ ,$$

are completely monotone (CM) [27,21]. Recall that a smooth function $f : [0, \infty) \to \mathbb{R}$ satisfying the following condition:

$$(-1)^n \frac{d^n}{ds^n} f(s) \geq 0, \quad s \geq 0, \quad n = 0, 1, 2, \ldots \tag{39}$$

The equivalence of (31) and (38) follows from the celebrated Bernstein theorem [27] which states that a function $f : [0, \infty) \to \mathbb{R}$ is completely monotone on $[0, \infty)$ if and only if it is the Laplace transform of a finite non-negative Borel measure μ on $[0, \infty)$, i.e. f is of the form

$$f(s) = \int_0^\infty e^{-st} d\mu(t). \tag{40}$$

In Ref. 21 we proved the following result

Theorem 3.1. *Let $W(s)$ be a function such that $1/[sW(s)]$ is CM. Then the set of functions*

$$\tilde{\kappa}_i(s) = -\frac{s}{a_i W(s) - 1}, \quad i = 1, 2, 3, \tag{41}$$

with $a_1, a_2, a_3 > 0$ such that

$$\frac{1}{s}\left(4 - \frac{1}{W(s)}\left[\frac{1}{a_1} + \frac{1}{a_2} + \frac{1}{a_3}\right]\right) \tag{42}$$

is CM, and

$$\frac{1}{a_1} + \frac{1}{a_2} \geq \frac{1}{a_3}, \quad \frac{1}{a_2} + \frac{1}{a_3} \geq \frac{1}{a_1}, \quad \frac{1}{a_3} + \frac{1}{a_1} \geq \frac{1}{a_2}, \tag{43}$$

define a legitimate memory kernel (49).

Let us observe that introducing $f(t) = 1/W(t)$ one finds

$$\tilde{\kappa}_i(s) = -\frac{s\tilde{f}(s)}{a_i - \tilde{f}(s)}, \quad i = 1, 2, 3, \tag{44}$$

and the condition (42) reduces to

$$0 \leq \int_0^t f(\tau)d\tau \leq 4\left(\frac{1}{a_1} + \frac{1}{a_2} + \frac{1}{a_3}\right)^{-1}. \tag{45}$$

Finally, defining $f_i(t) = f(t)/a_i$ one gets

$$\tilde{\kappa}_i(s) = -\frac{s\tilde{f}_i(s)}{1 - \tilde{f}_i(s)}, \quad i = 1, 2, 3, \tag{46}$$

which provides a generalization of (20).

4. Examples of legitimate kernels

Let us illustrate our result by a series of examples.

Example 4.1. The simplest example consists in taking $a_1 = a_2 = a_3 = a$. One easily finds that (43) are satisfied and

$$\tilde{\kappa}_i(s) = \tilde{\kappa}(s) = -\frac{s\tilde{f}(s)}{1 - \tilde{f}(s)}, \quad i = 1, 2, 3, \tag{47}$$

44

with

$$0 \leq \int_0^t f(\tau)d\tau \leq \frac{4}{3} \, . \tag{48}$$

In this *isotropic* case one has

$$K_t[\rho] = \frac{1}{2}\kappa(t) \sum_{i=1}^{3} (\sigma_i \rho \sigma_i - \rho) \, . \tag{49}$$

Example 4.2. Let a_1, a_2, a_3 satisfy (43) and

$$\widetilde{f}(s) = \frac{1}{(s+z_1)\dots(s+z_n)}, \tag{50}$$

with $z_i > 0$. If

$$\prod_{i=1}^{n} z_i \geq \frac{1}{4}\left(\frac{1}{a_1} + \frac{1}{a_2} + \frac{1}{a_3}\right), \tag{51}$$

then $\kappa_i(t)$ defined via (41) define a legitimate memory kernel[21]. One finds

$$\tilde{\kappa}_i(s) = -\frac{1}{a_i} \frac{s}{(s - s_1^{(i)})\dots(s - s_m^{(i)})}, \tag{52}$$

where $\{s_1^{(i)}, \dots, s_m^{(i)}\}$ are roots of the polynomial

$$P_i(s) = a_i(s+z_1)\dots(s+z_n) - 1,$$

for $i = 1, 2, 3$.

Example 4.3. Consider the simplest case of (50)

$$\widetilde{f}(s) = \frac{1}{s+z}, \tag{53}$$

with $z > 0$. One finds

$$\tilde{\kappa}_i(s) = -\frac{s}{a_i(s+z) - 1}, \tag{54}$$

and the inverse Laplace transform gives

$$\kappa_k(t) = -\frac{1}{z}\left(\delta(t) - \left[z - \frac{1}{a_k}\right]e^{-[z-\frac{1}{a_k}]t}\right). \tag{55}$$

Note, that if $a_k = 1/z$, then the dynamics is purely local. One easily finds for the eigenvalues of the dynamical map Λ_t

$$\lambda_k(t) = 1 - \frac{1}{za_k}(1 - e^{-zt}). \tag{56}$$

Interestingly, taking

$$a_1 = a_2 = \frac{1}{c} , \quad a_3 = \frac{1}{2c}. \tag{57}$$

One finds

$$\tilde{\kappa}_1(s) = \tilde{\kappa}_2(s) = -\frac{sc}{s+c}, \quad \tilde{\kappa}_3(s) = -2c,$$

and hence

$$\kappa_1(t) = \kappa_2(t) = -c\delta(t) + c^2 e^{-ct} , \quad \kappa_3(t) = -2c\delta(t)$$

which gives rise to the following memory kernel

$$K_t[\rho] = \frac{c}{2}\delta(t)[\sigma_1\rho\sigma_1 + \sigma_2\rho\sigma_2 - 2\rho] - \frac{c^2}{2}e^{-ct}[\sigma_3\rho\sigma_3 - \rho]. \tag{58}$$

In terms of the eigenvalues one has

$$\lambda_1(t) = \lambda_2(t) = \frac{1}{2}\left(1 + e^{-2ct}\right), \quad \lambda_3(t) = e^{-2ct}.$$

Interestingly, this evolution reproduces time-local description with

$$\gamma_1(t) = \gamma_2(t) = \frac{c}{2}, \quad \gamma_3(t) = -\frac{c}{2}\tanh(ct) \tag{59}$$

as discussed in [26].

5. Conclusions

In conclusion, we have analyzed quantum evolution of the qubit within memory kernel approach. It should be stressed that there is no general method to construct a legitimate kernel giving rise to completely positive dynamical map Λ_t. In the case of so-called random unitary evolution represented by a family of Pauli channel we derived a set of sufficient conditions summarized in the Theorem 3.1. This result enables one to construct a legitimate kernel using a set of positive numbers a_1, a_2, a_3 and a real function $f(t)$. Interestingly, numbers a_1, a_2, a_3 provide an analog of relaxation times within time-local approach [21]. Our analysis is illustrated by several simple examples. Surprisingly, many examples of qubit evolution analyzed in the literature fit our class of kernels. It would be interesting to generalize this result for higher dimensional systems.

Acknowledgments

This paper was partially supported by the National Science Center project 2015/17/B/ST2/02026. It is a pleasure to thank Professor Noboru Watanabe for his warm hospitality during QBIC 2015.

References

1. H.-P. Breuer and F. Petruccione, *The Theory of Open Quantum Systems* (Oxford Univ. Press, Oxford, 2007).
2. U. Weiss, *Quantum Dissipative Systems*, (World Scientific, Singapore, 2000).
3. R. Alicki and K. Lendi, *Quantum Dynamical Semigroups and Applications* (Springer, Berlin, 1987).
4. M. A. Nielsen and I. L. Chuang, *Quantum Computation and Quantum Information* (Cambridge Univ. Press, Cambridge, 2000).
5. R. Horodecki, P. Horodecki, M. Horodecki and K. Horodecki, Rev. Mod. Phys. **81**, 865 (2009).
6. V. Gorini, A. Kossakowski, and E. C. G. Sudarshan, J. Math. Phys. **17**, 821 (1976).
7. G. Lindblad, Comm. Math. Phys. **48**, 119 (1976).
8. S. Nakajima, Prog. Theor. Phys. **20**, 948 (1958); R. Zwanzig, J. Chem. Phys. **33**, 1338 (1960).
9. S. Chaturvedi and J. Shibata, Z. Phys. B **35**, 297 (1979); N. H. F. Shibata and Y. Takahashi, J. Stat. Phys. **17**, 171 (1977).
10. F. Haake, *Statistical Treatment of Open Systems by Generalized Master Equations*, (Springer Tracts in Modern Physics) vol. 66 (Berlin: Springer 1973).
11. S. M. Barnett and S. Stenholm, Phys. Rev. A **64**, 033808 (2001).
12. A. Shabani and D.A. Lidar, Phys. Rev. A **71**, 020101(R) (2005).
13. S. Maniscalco, Phys. Rev. A **72**, 024103 (2005); S. Maniscalco and F. Petruccione, Phys. Rev. A **73**, 012111 (2006).
14. A. A. Budini, Phys. Rev. A **69**, 042107 (2004).
15. J. Wilkie, Phys. Rev. E **62**, 8808 (2000); J. Wilkie and Yin Mei Wong, J. Phys. A: Math. Theor. **42**, 015006 (2009);
16. A. Kossakowski and R. Rebolledo, Open Syst. Inf. Dyn. **14**, 265 (2007); *ibid.* **16**, 259 (2009).
17. D. Chruściński and A. Kossakowski, EPL **97**, 20005 (2012).
18. H.-P. Breuer and B. Vacchini, Phys. Rev. Lett. **101** (2008) 140402; Phys. Rev. E **79**, 041147 (2009).
19. B. Vacchini, A. Smirne, E.-M. Laine, J. Piilo, and H.-P. Breuer, New J. Phys. **13**, 093004 (2011).
20. B. Vacchini, Phys. Rev. A **87**, 030101(R) (2013).
21. F. A. Wudarski, P. Należyty, G. Sarbicki, and D. Chruściński, Phys. Rev. A **91**, 042105 (2015).
22. D. Chruściński and A. Kossakowski, *Sufficient conditions for memory kernel master equation*, arXiv:1602.01642.
23. S. Lorenzo, F. Ciccarello, and G. M. Palma, *A class of exact memory-kernel master equations*, arXiv:1603.00248.
24. D. Chruściński and F. A. Wudarski, Phys. Lett. A **377**, 1425 (2013).
25. D. Chruściński, and F. A. Wudarski, Phys. Rev. A **91**, 012104 (2015).
26. E. Andersson, J. D. Cresser, and M. J. W. Hall, Phys. Rev. A **89**, 042120 (2014).
27. K. S. Miller, and S. G. Samko, Integr. Transf. and Spec. Funct. **12**, No 4, 389-402, (2001).

Quantum Bio-Informatics VI
© 2020 World Scientific Publishing Co. Pte. Ltd.
pp. 47–60

SYMMETRISED POSITION DISTRIBUTION OF QUANTUM STATES ON THE FOCK SPACE

K.-H. FICHTNER

University Jena, Institute of Applied Mathematics, 07743 Jena, Germany
E-mail: fichtner@mathematik.uni-jena.de

W. FREUDENBERG

Techn. University Cottbus, Dep. of Mathematics, 03013 Cottbus, Germany
E-mail: wolfgang.freudenberg@b-tu.de

It was shown in [4] that to each locally normal state on the symmetric Fock space one can associate a point process that can be interpreted as the position distribution of the state. The point process contains all information one can get by position measurements and is determined by the latter. In the present paper we deal with the problem whether one can obtain similar results also for fermion systems or for arbitrary states on the full Fock space. As an application we will investigate the position distribution of certain states introduced in [1].

1. Introduction

The state of a quantum particle in $\mathbb{R}^d$ is given by a density matrix (a normalised positive trace-class operator) on the Hilbert space $\mathfrak{h} = L^2(\mathbb{R}^d)$. Passing over to finite systems of particles we have to distinguish between bosonic and fermionic systems. Normal states of bosonic systems are given by density matrices on the symmetric Fock space $\Gamma^+(\mathfrak{h})$ while fermionic states are characterised by density matrices on the antisymmetric (or odd) Fock space $\Gamma^-(\mathfrak{h})$ over $\mathbb{R}^d$. In [4] we related to each normal bosonic state ω a point process Q_ω on $\mathbb{R}^d$ that can be interpreted as position distribution of ω. Hereby, $Q\omega$ is concentrated on finite configurations. In order to characterise infinite systems one may use positive normalised linear functionals ω on a suitably chosen C^*-sub algebra of the space of bounded linear operators on the symmetric Fock space $\Gamma^+(\mathfrak{h})$. In [4, Theorem 3.2] it was shown that also to those (so-called *locally normal*) states one can relate an infinite point process that can be interpreted as the position distribution of the state. Further, without any trouble all results in [4] may be transferred from $\mathbb{R}^d$

to an arbitrary Polish space G equipped with a locally finite measure μ.
A state of a quantum boson system never can be characterised solely by
its position distribution (unless the system would be a classical one). Besides the position distribution one needs still a special function (which was
called in [5, 6] conditional reduced density matrix) to determine the state
completely. The idea is to consider measurements that can be divided into
two parts - an application of a local observable to a finite subsystem and a
position measurement to the possibly infinite rest configuration. For details
we refer to [5, 6].

In the present paper we are concerned with the problem whether the results
obtained for bosonic systems can be transferred to systems of fermions or
to still more general states on the *full* Fock space. We will see that also for
fermionic systems the position distribution is a point process (describing
the positions of a *symmetric* particle system) though the wave function
of such systems will be *antisymmetric*. Generalisations to locally normal
states are straight forward. However, in the present paper we restrict our
considerations to finite quantum systems. For fermionic systems this will be
no restriction since because of the Pauli principle the position distribution
of a fermionic state always will be a finite point process. For the case
of a finite space G Bach and Zessin investigated in [1] different statistical
operators and certain point processes related to the corresponding states.
We will embed these results into our framework.

2. Position distribution

Let G be an arbitrary complete separable metric space and $\mathfrak{G}$ its σ-algebra
of Borel sets. Further, let μ be a fixed locally finite measure on $[G, \mathfrak{G}]$, i.e.
$\mu(B) < \infty$ for all bounded $B \in \mathfrak{G}$. Mainly, we are concerned with two
cases:

- $G = \mathbb{R}^d$, $d \geq 1$ and $\mu = \ell^d$ the Lebesgue measure on $\mathbb{R}^d$,
- $G = \{x_1, x_2, \ldots\}$ a countable set (finite or infinite) and $\mu = \sum_{k \geq 1} \delta_{x_k}$ the
counting measure on G.

Hereby, δ_x is the Dirac measure related to $x \in G$, i.e.

$$\delta_x(B) = \begin{cases} 1 & \text{if} \quad x \in B \\ 0 & \text{if} \quad x \notin B \end{cases} \qquad (B \in \mathfrak{G}). \qquad (1)$$

The basic space for our considerations will be the Hilbert space

$$\mathfrak{h} := L^2(G, \mu) = \left\{ \Psi : G \longrightarrow \mathbb{C}, \; \Psi \text{ measurable and } \int_G |\Psi(x)|^2 \mu(dx) < \infty \right\}.$$

The space of all bounded linear operators on $\mathfrak{h}$ we will denote by $\mathcal{L}(\mathfrak{h})$. Observe that in the discrete case mentioned above one has $\int_G \Psi(x)\mu(dx) = \sum_{k\geq 1} \Psi(x_k)$.

For measurable real-valued functions $f : G \longrightarrow \mathbb{R}$ we define the multiplication operator O_f on $\mathfrak{h}$ by

$$O_f \Psi(x) := f(x)\Psi(x) \qquad (x \in G, \; \Psi \in \mathfrak{h}). \tag{2}$$

Obviously, for bounded f it holds $O_f \in \mathcal{L}(\mathfrak{h})$. Especially, for $Y \in \mathfrak{G}$ the operator $O_Y := O_{\mathbb{1}_Y}$ measuring whether a particle belongs to the subset Y belongs to $\mathcal{L}(\mathfrak{h})$. Hereby, $\mathbb{1}_Y$ is the indicator function of the set Y, i.e. $\mathbb{1}_Y(x) = \delta_x(Y)$. All operators $O_f \in \mathcal{L}(\mathfrak{h})$ are self-adjoint.

Definition 2.1. Let ρ be a state on $\mathfrak{h} = L^2(G, \mu)$, i.e. ρ is a positive normalised trace-class operator on $\mathfrak{h}$. The unique probability measure Q_ρ on $[G, \mathfrak{G}]$ characterised by

$$\mathrm{Tr}(\rho O_f) = \int_G f(x)Q_\rho(dx) \quad (f : G \longrightarrow \mathbb{R} \text{ measurable and bounded}) \tag{3}$$

is called the *position distribution* of the state ρ.

Hereby, $\mathrm{Tr}(W)$ denotes the trace of an operator W. Existence and uniqueness of the probability measure Q_ρ follow easily from the properties of the trace.

The operators $O_f \in \mathcal{L}(\mathfrak{h})$ represent position measurements. Since $\mathrm{Tr}(\rho O_f)$ is the quantum mechanical expectation of the measurement O_f the relation (3) justifies to call Q_ρ the position distribution of the state ρ. Especially, for all $Y \in \mathfrak{G}$ we have $Q_\rho(Y) = \mathrm{Tr}(\rho O_Y)$ is the probability that the position of the quantum particle will be in the area Y.

3. The full Fock space

The (full) Fock space $\Gamma(\mathfrak{h})$ over $\mathfrak{h}$ is defined to be the Hilbert space completion of the direct sum of tensor products of copies of the underlying Hilbert space $\mathfrak{h}$, i.e.

$$\Gamma(\mathfrak{h}) = \overline{\bigoplus_{n=0}^{\infty} \mathfrak{h}^{n\otimes}}.$$

Hereby, $\mathfrak{h}^{0\otimes} = \mathbb{C}$ represents the vacuum. Because of the special choice $\mathfrak{h} = L^2(G, \mu)$ we may identify for $n \geq 1$ the tensor product $\mathfrak{h}^{n\otimes}$ with the space $L^2(G^n, \mu^{n\times})$ where $\mu^{n\times}$ is the n-th product measure of μ. The locally finite measure $\mu^{n\times}$ is a measure on $[G^n, \mathfrak{G}^n]$ with $\mathfrak{G}^n$ denoting the σ-algebra of Borel sets in G^n. Each function $\Psi \in \Gamma(\mathfrak{h})$ may be represented in the form $\Psi = (\Psi_n)_{n=0}^\infty$ with $\Psi_0 \in \mathbb{C}$, $\Psi_m \in L^2(G^m, \mu^{m\times})$ for $m \geq 1$ satisfying

$$\sum_{n=1}^\infty \|\Psi_n\|^2_{L^2(G^n, \mu^{n\times})} < \infty.$$

It is convenient to represent the Fock space as an L^2-space. To do this we set $\hat{G} := \bigcup_{n \geq 0} G^n$ with $G^0 = \{\hat{o}\}$ denoting the void configuration, and equip $\hat{G}$ with the σ-algebra $\hat{\mathfrak{G}}$ generated by $\bigcup_{n \geq 0} \mathfrak{G}^n$. The Fock space measure $\hat{F}_\mu$ is defined to be the measure on $[\hat{G}, \hat{\mathfrak{G}}]$ the restriction of which to G^n coincides with $\mu^{n\times}$, i.e. $\hat{F}_{\mu|_{G^n}} = \mu^{n\times}$ and $\hat{F}_\mu(\{\hat{o}\}) = 1$. For $\Psi = (\Psi_n)_{n=0}^\infty \in \Gamma(\mathfrak{h})$ one has

$$\Psi(x_1, \ldots, x_m) = \Psi_m(x_1, \ldots, x_m) \qquad (x_j \in G, \ j \in \{1, \ldots, m\}, \ m \in \mathbb{N}).$$

This way we may identify the Fock space $\Gamma(\mathfrak{h})$ with the L^2-space

$$\hat{\mathcal{H}} := L^2(\hat{G}, \hat{F}_\mu) = \left\{ \Psi = (\Psi_n)_{n=0}^\infty : \int_{\hat{G}} |\Psi(\varphi)|^2 \hat{F}_\mu(d\varphi) < \infty \right\}. \tag{4}$$

The scalar product in $\hat{\mathcal{H}}$ is given by

$$< \Psi, \Phi >_{\hat{\mathcal{H}}} := \overline{\Psi_0} \Phi_0 + \sum_{n=1}^\infty \int_{G^n} \overline{\Psi_n(\overline{x}^{(n)})} \Phi_n(\overline{x}^{(n)}) \mu^{n\times}(d[\overline{x}^{(n)}]) \tag{5}$$

where $\overline{\psi}$ denotes the complex conjugate of Ψ.

Remark 3.1. For each $n \geq 1$ we set

$$\hat{G}^n_s := \{(x_1, \ldots, x_n) : x_j \in G, \ x_j \neq x_k \text{ for } j \neq k, \ j, \ k \in \{1, \ldots, n\}\}.$$

The set $\hat{G}_s := \bigcup_{n=0}^\infty \hat{G}^n_s \subseteq \hat{G}$ is the set of *simple* configurations in $\hat{G}$ (configurations without multiple points). Observe that the Fock space measure $\hat{F}_\mu$ is concentrated on $\hat{G}_s$ if the measure μ on $[G, \mathfrak{G}]$ is diffuse (i. e. if $\mu(\{x\}) = 0$ for all $x \in G$). In this case a function $\Psi \in \hat{\mathcal{H}}$ is fully characterised by its restriction to $\hat{G}_s$.

Since we represented the Fock space as an L^2-space we can use Definition 2.1 to define the position distribution of a state on the Fock space.

Definition 3.1. Let ρ be a state on $\hat{\mathcal{H}} = L^2(\hat{G}, \hat{F}_\mu)$, i.e. ρ is a positive normalised trace-class operator on $\hat{\mathcal{H}}$. The unique probability measure Q_ρ on $[\hat{G}, \hat{\mathfrak{G}}]$ characterised by

$$\mathrm{Tr}(\rho O_f) = \int_{\hat{G}} f(\varphi) Q_\rho(d\varphi) \quad (f : \hat{G} \longrightarrow \mathbb{R} \text{ measurable and bounded}) \quad (6)$$

is called the *position distribution* of the state ρ.

Now, let ω be a pure state on $\hat{\mathcal{H}}$, i.e. there exists a (wave) function $\Psi \in \hat{\mathcal{H}}$, $\|\Psi\| = 1$ such that $\rho = |\Psi><\Psi| = <\Psi, \cdot> \Psi$ (ρ is the projection onto the one-dimensional subspace in $\hat{\mathcal{H}}$ generated by Ψ). Then for bounded measurable f we get

$$\mathrm{Tr}(\rho O_f) = <\Psi, O_f \Psi> = \int_{\hat{G}} f(\varphi) |\Psi(\varphi)|^2 \hat{F}_\mu(d\varphi).$$

We conclude that $Q_\rho \ll \hat{F}_\mu$ and $|\Psi|^2 = \frac{dQ_\rho}{d\hat{F}_\mu}$.

Immediately we may transfer this result to *normal* states on $\hat{\mathcal{H}}$, i.e. states ρ having the presentation

$$\rho = \sum_{k \geq 1} \alpha_k |\Psi_k><\Psi_k|. \quad (7)$$

Hereby, $(\Psi_k)_{k \geq 1}$ is a (finite or countable) orthonormal system in $\hat{\mathcal{H}}$, i. e. $\Psi_k \in \hat{\mathcal{H}}$, $\|\Psi_k\| = 1$, $\Psi_k \perp \Psi_l$ for $k \neq l$, $k, l \geq 1$, and $(\alpha_k)_{k \geq 1}$ is a probability distribution on $\mathbb{N}$ ($\alpha_k \geq 0$, $\sum_{n \geq 1} \alpha_n = 1$).

Proposition 3.1. *Let ρ be a normal state given in the form (7). Then*

$$Q_\rho \ll \hat{F}_\mu, \quad \frac{dQ_\rho}{d\hat{F}_\mu} = \sum_{k \geq 1} \alpha_k |\Psi_k|^2. \quad (8)$$

4. Bosonic and fermionic Fock space

A function $\Psi = (\Psi_n)_{n \geq 0} \in \hat{\mathcal{H}}$ is called *symmetric* resp. *anti-symmetric* if for all $n \geq 1$ and all permutations σ of $\{1, \ldots, n\}$ it holds

$$\Psi_n(x_1, \ldots, x_n) = \Psi_n(x_{\sigma(1)}, \ldots, x_{\sigma(n)}) \quad (x_1, \ldots, x_n \in G),$$

resp.

$$\Psi_n(x_1, \ldots, x_n) = \mathrm{sgn}(\sigma) \cdot \Psi_n(x_{\sigma(1)}, \ldots, x_{\sigma(n)}) \quad (x_1, \ldots, x_n \in G)$$

where sgn(σ) denotes the signature of the permutation σ.

Definition 4.1. By $\hat{\mathcal{H}}^+$ resp. $\hat{\mathcal{H}}^-$ we denote the subspaces of all symmetric resp. anti-symmetric functions from $\hat{\mathcal{H}}$. We will call $\hat{\mathcal{H}}^+$ the *bosonic or symmetric* Fock space and $\hat{\mathcal{H}}^-$ the *fermionic or anti-symmetric* Fock space. Bosonic resp. fermionic states are states of the form (7) with $\Psi_k \in \hat{\mathcal{H}}^+$ resp. $\Psi_k \in \hat{\mathcal{H}}^-$.

Observe that for $\Psi \in \hat{\mathcal{H}}^+$ and $\Psi \in \hat{\mathcal{H}}^-$ the function $|\Psi|^2$ is symmetric. Consequently, from Proposition 3.1 we conclude that the density $dQ_\rho/d\hat{F}_\mu$ both for bosonic and fermionic states are symmetric functions. Summarising we obtain

Proposition 4.1.

(a) *Denote by $\mathcal{F}$ the set of all bounded measurable and symmetric function $f : \hat{G} \longrightarrow \mathbb{R}$. The spaces $\hat{\mathcal{H}}^+$ and $\hat{\mathcal{H}}^-$ are invariant with respect to O_f for $f \in \mathcal{F}$, i.e.*

$$\{O_f \Psi : \Psi \in \hat{\mathcal{H}}^\pm\} \subseteq \hat{\mathcal{H}}^\pm \qquad (f \in \mathcal{F}).$$

(b) *The position distribution Q_ρ of a bosonic or fermionic state is determined by*

$$\mathrm{Tr}(\rho O_f) = \int_{\hat{G}} f(\varphi) Q_\rho(d\varphi) \qquad (f \in \mathcal{F}),$$

i. e. the position measurements O_f with symmetric f characterise the position distribution completely.

(c) *The position distribution of a bosonic or fermionic state is absolutely continuous with respect to the Fock space measure $\hat{F}_\mu$ with symmetric Radon-Nikodym derivative (cf. (8)).*

Remark 4.1. Generalisations of Proposition 4.1 to locally normal states of bosonic states are given in [4, 5].

Example 4.1. Each probability measure Q on $[\hat{G}, \hat{\mathfrak{S}}]$ being absolutely continuous with respect to $\hat{F}_\mu$ with symmetric Radon-Nikodym derivative g gives rise to a bosonic state ρ with position distribution $Q_\rho = Q$. The state

$$\rho := |\Psi>< \Psi|, \qquad \Psi := \sqrt{g}$$

is a pure Bosonic state with position distribution Q. But also for arbitrary $f : \hat{G} \longrightarrow \mathbb{R}$ the pure state $\rho := |\Psi>< \Psi|$ with $\Psi = e^{if}\sqrt{g}$ will have the same position distribution Q.

Example 4.2. We consider the case $G = \mathbb{R}^1$ equipped with the Lebesgue measure ℓ. For each $n \in \mathbb{N}$ and all $\overline{x}^{(n)} = (x_1, \ldots, x_n) \in \hat{G}_s^n$ there exists exactly one permutation π of $\{1, \ldots, n\}$ (depending on $\overline{x}^{(n)}$) such that $x_{\pi(1)} < \cdots < x_{\pi(n)}$. The signature of this permutation we will denote by $\mathrm{sgn}(\overline{x}^{(n)})$. Let $\Psi = (\Psi_n)_{n=0}^\infty \in \hat{\mathcal{H}}^+$, $\|\Psi\| = 1$ be a normalised symmetric wave function. For $\overline{x}^{(m)}) \in \hat{G}_s^m$, $m \geq 1$ we set

$$\Pi_m \Psi_m(\overline{x}^{(m)}) := \mathrm{sgn}(\overline{x}^{(m)}) \cdot \Psi_m(\overline{x}^{(m)}). \tag{9}$$

From Remark 3.1 we conclude that the Fock space measure $\hat{F}_\ell$ is concentrated on $\hat{G}_s$. Thus, $\Pi\Psi := (\Pi_n \Psi_n)_{n=0}^\infty$ is fully determined by Example 4.1. It is easy to see that $\Pi\Psi \in \hat{\mathcal{H}}^-$, $\|\Pi\Psi\| = 1$ and $|\Pi\Psi|^2 = |\Psi|^2$ $\hat{F}_\ell - a.e.$ The state $\rho = |\Psi><\Psi|$ is bosonic, the state $\hat{\rho} = |\Pi\Psi><\Pi\Psi|$ is a fermionic one. However, the position distribution of both states coincide: $Q_\rho = Q_{\hat{\rho}}$. To each probability measure on Q on $[\hat{G}, \hat{\mathfrak{G}}]$ with symmetric Radon-Nikodym derivative with respect to $\hat{F}_\mu$ we may associate as well a bosonic as also a fermionic state having this probability measure as position distribution.

5. Point processes

Point processes as distributions of random counting measures represent a useful tool to describe random symmetric point configurations. Since the position distributions of bosonic or fermionic states live on symmetric point configurations we transfer above notions into the language of counting measures.

Observe that a *symmetric* configuration $(x_1, \ldots, x_n)$ of n points from G is fully characterized by the *counting measure* $\varphi = \delta_{x_1} + \cdots + \delta_{x_n}$ where δ_x is the Dirac measure defined by (1). For each $n \in \mathbb{N}$ we denote by $M_n(G)$ the set of all symmetric n-particle configurations from G, i. e.

$$M_n(G) = \{\varphi = \delta_{x_1} + \cdots + \delta_{x_n} : x_1, \ldots, x_n \in G\}.$$

The empty configuration $\mathfrak{o}$ is described by the null measure on G, i. e. we have $\mathfrak{o}(G) = 0$, and we set

$$M(G) = \bigcup_{n=0}^\infty M_n(G) \tag{10}$$

with $M_0(G) = \{\mathfrak{o}\}$. We equip $M(G)$ with its canonical σ-algebra $\mathfrak{M}(G)$ — the smallest σ-algebra containing all sets of the form $\{\varphi \in M(G) : \varphi(K) = n\}$, $K \in \mathfrak{G}$, $n \in \mathbb{N}$. Observe that $\varphi(K) = n$ means that the configuration φ has exactly n points in the subset K of the phase space G.

Definition 5.1. A point process on G is a probability measure on the space $[M(G), \mathfrak{M}(G)]$ of counting measures.

The mapping $\xi : \hat{G} \longrightarrow M(G)$

$$\xi(\overline{x}^{(n)}) := \varphi_{\overline{x}^{(n)}} := \sum_{k=1}^{n} \delta_{x_k} \qquad (\overline{x}^{(n)} = (x_1, \ldots, x_n) \in G^n, \ n \in \mathbb{N}) \quad (11)$$

with $\xi(\hat{o}) = o$ maps $\hat{G}$ onto the set $M(G)$. Further, the mapping

$$J(\Psi)(\overline{x}) := \frac{\varphi_{\overline{x}}(G)!}{\prod_{y \in G}(\varphi_{\overline{x}}(\{y\}))!} \Psi(\xi(\overline{x})) \qquad (\overline{x} \in \hat{G}, \ \Psi : M(G) \longrightarrow \mathbb{C}) \quad (12)$$

allows to identify functions Ψ on $M(G)$ with symmetric functions on $\hat{G}$. (Observe that the product in the denominator is always a finite one.) We will replace the bosonic Fock space $\hat{\mathcal{H}}^+$ by an isomorphic L^2-space on $[M(G), \mathfrak{M}(G)]$. The measure $\hat{F}_\mu$ on $[\hat{G}, \hat{\mathfrak{G}}]$ we replace by the measure F_μ on $[M(G), \mathfrak{M}(G)]$ given by

$$F_\mu(Y) := \mathbb{1}_Y(o) + \sum_{n \geq 1} \frac{1}{n!} \int_{G^n} \mathbb{1}_Y\left(\xi(\overline{x}^{(n)})\right) \mu^n(d\overline{x}^{(n)}) \qquad (Y \in \mathfrak{M}(G)).$$

$$(13)$$

The relation between $\hat{F}_\mu$ and F_μ is given by

$$\int_{M(G)} \Psi(\varphi) F_\mu(d\varphi) = \int_{\hat{G}} J^{-1}(\Psi)(\xi(\overline{x})) \hat{F}_\mu(d\overline{x})$$

$$= \int_{\hat{G}} \frac{\prod_{y \in G}(\varphi_{\overline{x}}(\{y\}))!}{\varphi_{\overline{x}}(G)!} \Psi(\xi(\overline{x})) \hat{F}_\mu(d\overline{x}). \quad (14)$$

We denote by $\mathcal{H}^+ := L^2(M(G), \mathfrak{M}(G), F_\mu)$ the space of square integrable complex-valued functions on $M(G)$, i. e.

$$\mathcal{H}^+ := \left\{ \Psi : M(G) \longrightarrow \mathbb{C} : \int_{M(G)} |\Psi(\varphi)|^2 F_\mu(d\varphi) < \infty \right\} \quad (15)$$

The scalar product in $\mathcal{H}^+$ is given by

$$\langle \Psi, \Phi \rangle_{\mathcal{H}^+} := \int_{M(G)} \overline{\Psi(\varphi)} \cdot \Phi(\varphi) F_\mu(d\varphi) = \int_{\hat{G}} J(\overline{\Psi} \cdot \Phi)(\overline{x}) \hat{F}_\mu(d\overline{x}).$$

On the other hand, since for $\Psi, \ \Phi \in \hat{\mathcal{H}}^+$ as well as for $\Psi, \ \Phi \in \hat{\mathcal{H}}^-$ we have $\overline{\Psi} \cdot \Phi \in \hat{\mathcal{H}}^+$ we can express in both cases the scalar product in the space

$\mathcal{H}^+$. For Ψ, $\Phi \in \hat{\mathcal{H}}^\pm$ one has

$$\langle \Psi, \Phi \rangle_{\hat{\mathcal{H}}} := \int_{\hat{G}} \overline{\Psi(\overline{x})} \Phi(\overline{x}) \hat{F}_\mu(d\overline{x}) = \int_{M(G)} J^{-1}(\overline{\Psi} \cdot \Phi)(\varphi) F_\mu(d\varphi). \qquad (16)$$

Proposition 5.1. *The spaces $\hat{\mathcal{H}}^+$ and $\mathcal{H}^+$ are isomorphic. The isomorphism is given by the unitary operator $U : \mathcal{H}^+ \longrightarrow \hat{\mathcal{H}}^+$*

$$U\Psi(\overline{x}) := \sqrt{\frac{\varphi_{\overline{x}}(G)!}{\prod_{y \in G}(\varphi_{\overline{x}}(\{y\})!)}} \, \Psi(\varphi_{\overline{x}}) \qquad (\Psi \in \mathcal{H}^+, \overline{x} \in \hat{G}). \qquad (17)$$

Since $\hat{\mathcal{H}}^+$ and $\mathcal{H}^+$ are isomorphic also $\mathcal{H}^+$ will be called *bosonic* or *symmetric Fock space over G*. For details we refer to [4] or to [7] where a similar definition of the symmetric Fock space was given.

Observe that $\hat{F}_\mu$ is concentrated on the set $\hat{G}_s$ of simple configurations (without multiple points) if and only if F_μ is concentrated on the set

$$M(G)_s := \xi(\hat{G}_s) = \{\varphi \in M(G) : \varphi(\{x\}) \le 1 \text{ for all } x \in G\}$$

of simple counting measures. F_μ is concentrated on the set $M(G)_s$ if μ a diffuse measure on G. But also in the case of Ψ, $\Phi \in \hat{\mathcal{H}}^-$ the function $J^{-1}(\overline{\Psi} \cdot \Phi)$ will be concentrated on $M(G)_s$. In this case in the above formulae the term $\prod_{y \in G}(\varphi_{\overline{x}}(\{y\})!)$ is equal to 1, and e. g. (17) can be replaced by

$$U\Psi(\overline{x}) := \sqrt{n!} \, \Psi(\varphi_{\overline{x}}) \qquad (\Psi \in \mathcal{H}^+, \overline{x} \in G^n, \, n \in \mathbb{N}) \qquad (18)$$

and (16) reduces to

$$\langle \Psi, \Phi \rangle_{\hat{\mathcal{H}}} := \int_{\hat{C}} \overline{\Psi(\overline{x})} \Phi(\overline{x}) \hat{F}_\mu(d\overline{x}) = \int_{M(G)_s} \frac{1}{\varphi(G)!} \cdot (\overline{\Psi} \cdot \Phi)(\varphi) F_\mu(d\varphi). \qquad (19)$$

Proposition 5.2. *If μ is a finite diffuse measure on $[G, \mathfrak{G}]$ then F_μ is a finite measure and*

$$F_\mu(M(G)) = 1 + \sum_{n \ge 1} \frac{\mu^n(G^n)}{n!} = \sum_{n \ge 0} \frac{(\mu(G))^n}{n!} = e^{\mu(G)} < \infty.$$

Moreover, $e^{-\mu(G)} F_\mu$ is the Poisson point process with intensity measure μ.

To each normal state ρ on $\hat{\mathcal{H}}^+$ or $\hat{\mathcal{H}}^-$ we related its position distribution Q_ρ having the property (8) and being determined by $\mathrm{Tr}(\rho O_f)$ for $f \in \mathcal{F}$. All these notions we may transfer easily to the language of point processes.

Proposition 5.3. *Let ρ be a normal state on $\hat{\mathcal{H}}^+$ or $\hat{\mathcal{H}}^-$ with position distribution Q_ρ. Then $P_\rho := Q_\rho \circ \xi^{-1}$ is a point process on G with the properties:*

$$P_\rho \ll F_\mu \tag{20}$$

$$\int_{M(G)} g(\varphi) P_\rho(d\varphi) = \mathrm{Tr}(\rho O_{g \circ \xi}) \qquad (f \in \mathcal{F}) \tag{21}$$

In a series of papers starting with [3] we discussed the position distributions of several normal and locally normal states of boson systems. For instance in [2] it is shown that the position distribution of the ideal Bose gas is an infinitely divisible point process having a characteristic clustering representation. Further, of special interest are so-called coherent states describing boson systems where all particles are in the "one-particle state". These states are of the form $\rho = |\Psi><\Psi|$ where $\Psi \in \mathcal{H}^+$ is an exponential vector, i.e.

$$\Psi(\bar{x}) = g(x_1) \cdot \ldots \cdot g(x_n), \qquad g \in \mathfrak{h}, \ \bar{x} \in G^n, \ n \in \mathbb{N}.$$

The position distribution of such a coherent state is a Poisson process with intensity measure $\Lambda = \int |g|^2 d\mu$ (cf. [3, 4, 5]). Example 6.2 below is a special case of the position distribution of such a coherent state. Obviously, one may consider also mixtures of such states and pass over to locally normal states.

6. States on the Fock space with countable G

Let $G = \{x_1, x_2, \ldots\}$ be countable. Then we have

$$F_\mu(\{\varphi\}) = \prod_{x \in \varphi} \mu(\{x\})^{\varphi(\{x\})} \qquad (\varphi \in M(G))$$

where $x \in \varphi$ means $\varphi(\{x\}) \geq 1$. Remark that even for infinite G the product is a finite one. In the sequel we will be interested in the special case of μ being the counting measure on G, i.e. $\mu = \delta_{x_1} + \delta_{x_2} + \ldots$. Then F_μ is the counting measure on $M(G)$ and $\hat{F}_\mu$ the counting measure on $\hat{G}$. Especially, if G is finite, say $\mu(G) = m$ then

$$F_\mu(M_n(G)) = \binom{n + m - 1}{n} \tag{22}$$

$$F_\mu(M_n(G) \cap M(G)_s) = \binom{m}{n} \qquad (n \leq m) \tag{23}$$

According to Proposition 5.3 to each state ρ on $\hat{\mathcal{H}}$ we can relate its position distribution Q_ρ on $\hat{G}$ and for states on $\hat{\mathcal{H}}^\pm$ the corresponding point process $P_\rho := Q_\rho \circ \xi^{-1}$. The latter can be calculated directly from

$$\int_{M(G)} g(\varphi) P_\rho(d\varphi) = \mathrm{Tr}(\rho O_{g \circ \xi}) \qquad (g : M(G) \longrightarrow \mathbb{R}).$$

The point process P_ρ is absolutely continuous with respect to F_μ and its density is given by $p(\varphi) = P_\mu(\{\varphi\})$.

If the operator $\tilde{\rho}$ is a positive trace operator with finite trace (but not necessarily normalised) we obtain in the same way as above finite measures $Q_{\tilde{\rho}}$ on $\hat{G}$ resp. $Q_{\tilde{\rho}} \circ \xi^{-1}$ on $M(G)$. The normalisation of $Q_{\tilde{\rho}} \circ \xi^{-1}$ leads to a point process being the position distribution of the normalised operator ρ. More precisely, if $\tilde{\rho}$ is a positive trace class operator with $\mathrm{Tr}(\tilde{\rho}) < \infty$ then the normalised operator $\rho := \frac{\tilde{\rho}}{\mathrm{Tr}(\tilde{\rho})}$ is a state on the Fock space $\hat{\mathcal{H}}$ and it is easily checked that

$$Q_\rho = \frac{Q_{\tilde{\rho}}}{Q_{\tilde{\rho}}(\hat{G})} \qquad \text{and} \qquad Q_\rho \circ \xi^{-1} = \frac{Q_{\tilde{\rho}} \circ \xi^{-1}}{Q_{\tilde{\rho}} \circ \xi^{-1}(M(G))}$$

are the position distributions of ρ on $\hat{G}$ resp. $M(G)$. In the sequel we restrict ourselves to states ρ on $\hat{\mathcal{H}}$ being the normalisation of operators $\tilde{\rho}$ that can be written as the direct sum of operators $\tilde{\rho}_n$ on the n-particle spaces $L^2(G^n, \mu^{n\times})$, i.e.

$$\rho := \frac{\tilde{\rho}}{\mathrm{Tr}(\tilde{\rho})}, \qquad \tilde{\rho} := \bigoplus_{n=0}^{\infty} \tilde{\rho}_n. \qquad (24)$$

Bach and Zessin investigated in [1] for of a finite space G different statistical operators and related certain point processes to the corresponding states. We will embed now these results into our framework. The finite set G we replace by a countable one.

We set $G = \{x_1, x_2, \ldots\}$ and consider the space $\mathfrak{h} = L^2(G, \mu)$ with $\mu = \delta_{x_1} + \delta_{x_2} + \ldots$ being be the counting measure on G. The sequence $(\mathbb{1}_{x_n})_{n=1}^\infty$ establishes an orthonormal basis in $\mathfrak{h}$. Analogously, $(\mathbb{1}_{\bar{x}})_{\bar{x} \in \hat{G}}$ is an orthonormal basis in $\hat{\mathcal{H}} = L^2(\hat{G}, \hat{F}_\mu)$. For arbitrary states ρ on $\hat{\mathcal{H}}^\pm$ and $\bar{y} \in \hat{G}$, $\varphi \in M(G)$ one has

$$Q_\rho(\{\bar{y}\}) = \mathrm{Tr}(\rho O_{\mathbb{1}_{\bar{y}}}), \qquad P_\rho(\{\varphi\}) = Q_\rho \circ \xi^{-1}(\{\varphi\})) = \mathrm{Tr}(\rho O_{\mathbb{1}_{\xi^{-1}(\varphi)}}).$$

Now let τ be a state on $\mathfrak{h}$ of the form

$$\tau = \sum_{x \in G} \lambda(x) \langle \mathbb{1}_x, \cdot \rangle \, \mathbb{1}_x, \qquad \sum_{x \in G} \lambda(x) = 1, \qquad \lambda(x) \geq 0. \tag{25}$$

In the sequel we restrict ourselves to states ρ on $\hat{\mathcal{H}}^{\pm}$ being of the form $\rho := \bigoplus_{n=0}^{\infty} \rho_n$ where for each $n \in \mathbb{N}$ the operator ρ_n is a state

Example 6.1. For fixed $m \in \mathbb{N}$ consider the state $\rho^{(m)} := \bigoplus_{n=0}^{\infty} \tilde{\rho}_n$ with

$$\tilde{\rho}_m = \tau^{m\otimes}, \qquad \tilde{\rho}_n = 0 \qquad (n \neq m). \tag{26}$$

Obviously, for $\overline{y} \notin G^m$ one has $Q_{\rho^{(m)}}(\{\overline{y}\}) = 0$, and for $\overline{y} \in G^m$ we obtain

$$Q_{\rho^{(m)}}(\{\overline{y}\}) = \mathrm{Tr}(\rho^{(m)} O_{\mathbf{1}_{\overline{y}}}) = \mathrm{Tr}(\rho^{(m)} \mathbb{1}_{\overline{y}}) = \sum_{\overline{x} \in G^m} \langle \mathbb{1}_{\overline{x}}, \tau^{m\otimes} \mathbb{1}_{\overline{y}} \mathbb{1}_{\overline{x}} \rangle$$

$$= \langle \mathbb{1}_{\overline{y}}, \tau^{m\otimes} \mathbb{1}_{\overline{y}} \rangle = \prod_{k=1}^{m} \lambda(y_k) = \prod_{x \in G} \lambda(x)^{\varphi_{\overline{y}}(\{x\})}$$

Using (14) we obtain for arbitrary function h on $M(G)$

$$\int_{M(G)} P_{\rho^{(m)}}(\{\varphi\}) h(\varphi) F_{\mu}(d\varphi)$$

$$= \sum_{\varphi \in M(G)} P_{\rho^{(m)}}(\{\varphi\}) h(\varphi) = \sum_{\overline{y} \in \hat{G}} Q_{\rho^{(m)}}(\{\overline{y}\}) h(\overline{y}) = \int_{\hat{G}} Q_{\rho^{(m)}}(\{\overline{y}\}) h(\overline{y}) \hat{F}_{\mu}(d\overline{y})$$

$$= \int_{M(G)} \mathbb{1}_{M_n(G)}(\varphi) \prod_{x \in G} \lambda(x)^{\varphi(\{x\})} \frac{\varphi(G)!}{\prod_{y \in \varphi} \varphi(\{y\})!} h(\varphi) F_{\mu}(d\varphi)$$

Consequently, we get for all $\varphi \in M(G)$

$$P_{\rho^{(m)}}(\{\varphi\}) = Q_{\rho^{(m)}} \circ \xi^{-1}(\{\varphi\}) = \mathbb{1}_{M_n(G)}(\varphi) \prod_{x \in G} \lambda(x)^{\varphi(\{x\})} \frac{\varphi(G)!}{\prod_{y \in \varphi} \varphi(\{y\})!}. \tag{27}$$

In [1] the point process $Q_{\rho^{(m)}}$ is called *Maxwell Boltzmann process*.

Example 6.2. Let τ be as in (25) and $\rho := \sum_{m=1}^{\infty} \alpha_m \rho^{(m)}$ with $(\alpha_m)_{m=0}^{\infty}$ being a Poisson distribution with parameter $\nu > 0$ and $\rho^{(m)}$ defined by (26) as in Example 6.1. Then it is easy to check that $\rho^{(m)} = Q_{\rho^{(m)}} \circ \xi^{-1}$ is a *Poisson process* with intensity measure $\nu \cdot \lambda$.

Example 6.3. We consider a state $\rho = \tilde{\rho}/\mathrm{Tr}(\tilde{\rho})$ of the form (24) with

$$\tilde{\rho}_n = \Pi_+ \tau^{n \otimes} \Pi_+ \qquad (n \geq 0) \tag{28}$$

with $\Pi_+ = \bigoplus_{n=0}^{\infty} \Pi_{+,n}$ denoting the operator of symmetrization, i.e.

$$\Pi_{+,n} \mathbb{1}_{\overline{x}} = \frac{1}{n!} \sum_{\sigma} \mathbb{1}_{(x_{\sigma(1)}, \ldots, x_{\sigma(n)})} \qquad (\overline{x} \in G^n, \; n \geq 1)$$

where the sum runs over all permutations σ of $\{1, \ldots, n\}$. One may prove the following statement.

Proposition 6.1. *Assume $\lambda(x) < 1$ for all $x \in G$. The position distribution P_ρ of the (normalised) state ρ is given by*

$$P_\rho(\{\varphi\}) = Q_\rho \circ \xi^{-1}(\{\varphi\}) = \frac{F_\mu(\{\varphi\})}{F_\mu(M(G))} = \frac{\prod\limits_{x \in G} \lambda(x)^{\varphi(\{x\})}}{\sum\limits_{\varphi \in M(G)} \prod\limits_{x \in G} \lambda(x)^{\varphi(\{x\})}} \tag{29}$$

The assumption $\lambda(x) < 1$ is necessary to ensure $\mathrm{Tr}(\tilde{\rho}) < \infty$. The above point process is called in [1] *Bose Einstein process*.

Example 6.4. In analogy to the above example we consider a state $\rho = \tilde{\rho}/\mathrm{Tr}(\tilde{\rho})$ of the form (24) with

$$\tilde{\rho}_n = \Pi_- \tau^{n \otimes} \Pi_- \qquad (n \geq 0) \tag{30}$$

with $\Pi_- = \bigoplus_{n=0}^{\infty} \Pi_{-,n}$ denoting the operator of anti-symmetrization, i.e.

$$\Pi_{-,n} \mathbb{1}_{\overline{x}} = \frac{1}{n!} \sum_{\sigma} \mathrm{sgn}(\sigma) \mathbb{1}_{(x_{\sigma(1)}, \ldots, x_{\sigma(n)})} \qquad (\overline{x} \in G^n, \; n \geq 1)$$

where the sum runs over all permutations σ of $\{1, \ldots, n\}$ and $\mathrm{sgn}(\sigma)$ denotes the signature of the permutation σ. We get the following result

Proposition 6.2. *The position distribution P_ρ of the (normalised) state ρ is concentrated on the set of simple configurations $M(G)_s$, and is given by*

$$P_\rho(\{\varphi\}) = \frac{F_\mu(\{\varphi\})}{F_\mu(M(G)_s)} = \frac{\prod\limits_{x \in \varphi} \lambda(x)}{\sum\limits_{\varphi \in M(G)_s} \prod\limits_{x \in \varphi} \lambda(x)} \qquad (\varphi \in M(G)_s). \tag{31}$$

This type of point processes was called in [1] *Fermi Dirac process*.

References

1. A. Bach and H. Zessin. The particle structure of the bose and fermi gas. Technical report, University Bielefeld, 2013. 16 pages.
2. K.-H. Fichtner. On the position distribution of the ideal Bose gas. *Math. Nachr.*, 151:59 – 67, 1991.
3. K.-H. Fichtner and W. Freudenberg. On a probabilistic model of infinite quantum mechanical particle systems. *Math. Nachr.*, 121:171—210, 1985.
4. K.-H. Fichtner and W. Freudenberg. Point processes and the position distribution of infinite boson systems. *J. Stat. Phys.*, 47:959—978, 1987.
5. K.-H. Fichtner and W. Freudenberg. Characterization of states of infinite boson systems I. On the construction of states of boson systems. *Commun. Math. Phys.*, 137:315—357, 1991.
6. W. Freudenberg. Characterization of states of infinite boson systems II. On the existence of the conditional reduced density matrix. *Commun. Math. Phys.*, 137:461—472, 1991.
7. H. Maassen. Quantum Markov processes on Fock space described by integral kernels. In L. Accardi and W. von Waldenfels, editors, *Quantum Probability and Applications II*, volume 1136 of *Lecture Notes in Mathematics*, pages 361–374, Berlin, Heidelberg, New York, 1985. Springer.

Quantum Bio-Informatics VI
© 2020 World Scientific Publishing Co. Pte. Ltd.
pp. 61–71

NOTE ON QUANTUM ALGORITHM BASED ON ADAPTIVE DYNAMICS

SATOSHI IRIYAMA

Department of Information Sciences, Tokyo University of Science, Noda City, Chiba 278-8510, Japan
E-mail: iriyama@is.noda.tus.ac.jp

Dedicated to Professor Masanori Ohya

1. Introduction

In 1999, Ohya and Volovich showed that an NP complete problem can be solved by quantum algorithm in polynomial time of input length[1,2]. This algorithm is the essential idea to construct efficient quantum algorithm in our research group[3]. We applied this method not only to the mathematical problems, but also studies of the bio-informatics where the problems are described by the complex systems. Ohya, in 2008, proposed the notion of Adaptive Dynamics (AD) which is a mathematical framework to treat with complex systems such as Chaos, algorithm, quantum physics, bio-informatics, psychological phenomena and so on[4]. In this paper, I explain our research of quantum algorithm based on AD reviewing our result.

1.1. *adaptive dynamics*

Ohya introduced Adaptive Dynamics (AD) and it is appeared in series of many papers[4]. The AD indicates how to construct a mathematical model to treat with a complex phenomena. According to the aspects of AD, we can study the subjects efficiently with unified idea. The AD has the following two aspects:

- Observable-adaptive
  (1) Measurement depends on how to see an observable to be measured
  (2) The interaction between two systems depends on how a fixed observable exists

- State-adaptive
  (1) Measurement depends on how the state to be used exists
  (2) The correlation between two systems interaction depends on how the state of at least one of the systems at one instant exists

The notion of observable adaptive is applied to the study of understanding chaos and violation of Bell's inequality as examples. The notion of state adaptive is used to construct an efficient quantum algorithm solving NP complete problems. There are many applications using the AD such as mathematical analysis of chaos[3], decision making process for Prisoner's Dilemma game[5], Glucose effect on E-coli growth[6], and context dependent systems breaking the probability law[7].

2. Quantum Algorithm

A quantum algorithm is not usually more efficient than classical one, for example, Shor's factorization algorithm has same success probability as classical one[8]. Our subjects of quantum algorithm are classified as follows

(1) Halting problem
The halting problem indicates the computational limitation of classical algorithm where the algorithm is essentially restricted. The quantum version of halting problem is discussed in [9]. We showed that the halting probability of quantum algorithm can be estimated without its simulation time.

(2) Quantum Turing machine
Using the notion of state adaptive, we gave a mathematical model of quantum Turing machine[10]. The configuration of quantum Turing machine is described by the density operator on the Hilbert space, and the transition function is given by the completely positive channel. The precise definitions and main results will be given the next subsection.

(3) Hard problems
The one of NP complete problems is solved by quantum algorithm in a polynomial time[1,2,11]. An NP-hard problem is strictly harder problem than an NP complete problem. The most general type of searching problems which does not assume the existence of answer belongs to the class NP-hard. We constructed a quantum algorithm to solve it in polynomial time of the logarithm of searching space

size. The essential idea is that the quantum algorithm can be given by the combinations of quantum algorithm[12].

(4) Applications for bio-informatics

The alignment for amino acid sequence is a fundamental operation to study bio-informatics. Alignment for two sequences is called pairwise alignment. The MTRAP alignment is one of the fastest pairwise alignment method based on the idea of entanglement[13]. This classical algorithm shows that mathematics of quantum mechanics is useful to increase efficiency of algorithm.

2.1. *generalized Turing machine*

Based on State-AD, we defined a generalized Turing machine (GTM) including classical and quantum Turing machine[14]. The GTM is given by

$$M_{gq} = (Q, \Sigma, \mathcal{H}, \Lambda_\delta)$$

where

- Q: a set of states
- Σ: a set of alphabets
- $\mathcal{H}$: a Hilbert space spaned by a computational basis
- Λ_δ: a transition channel(program) on $\mathfrak{S}(\mathcal{H})$

Here, we define the transition function

$$\delta_1 : \mathbb{R} \times Q \times \Sigma \times Q \times \Sigma \times Q \times \Sigma \times \{0, \pm 1\} \times Q \times \Sigma \times \{0, \pm 1\} \to \mathbb{C}.$$

A quantum transition function is given by a quantum channel

$$\Lambda_{\delta_1} : \mathfrak{S}(\mathcal{H}) \to \mathfrak{S}(\mathcal{H}),$$

satisfying the following condition.

Definition 1. Λ_{δ_1} is called a quantum transition channel if there exists a transition function δ_1 such that for any quantum configuration $\rho = \sum_k \lambda_k |\psi_k\rangle \langle\psi_k|, \ |\psi_k\rangle = \sum_l \alpha_{k,l} |q_{k,l}, A_{k,l}, i_{k,l}\rangle, \ \sum_k \lambda_k = 1, \forall \lambda_k \geq 0,$ $\sum_l |\alpha_{k,l}|^2 = 1, \forall \alpha_{k,l} \in \mathbb{C}$ it holds

$$\Lambda_{\delta_1}(\rho) \equiv \sum_{k,l,m,n,p,b,d,p',b',d'} \delta_1\left(\lambda_k, q_{k,l}, A_{k,l}(i_{k,l}), q_{m,n}, A_{m,n}(i_{m,n}), p, b, d, p', b', d'\right)$$
$$\times |p, B, i_{k,l} + d\rangle \langle p', B', i_{m,n} + d'|$$

$$B\left(j\right)=\begin{cases} b & j=i_{k,l} \\ A_{k,l}\left(j\right) & \text{otherwise} \end{cases}$$

$$B'\left(j\right)=\begin{cases} b' & j=i_{m,n} \\ A_{m,n}\left(j\right) & \text{otherwise} \end{cases}$$

so that the RHS is a state.

Definition 2. $M_{gq} = (Q, \Sigma, \mathcal{H}, \Lambda_\delta)$ is called a LQTM (Linear Quantum Turing Machine) if there exists a transition function

$$\delta_2 : Q \times \Sigma \times Q \times \Sigma \times Q \times \Sigma \times \{0, \pm1\} \times Q \times \Sigma \times \{0, \pm1\} \to \mathbb{C}$$

such that for any quantum configuration ρ_k, Λ_{δ_2} is written as

$$\Lambda_{\delta_2}\left(\rho_k\right) \equiv \sum_{k,l,m,n,p,b,d,p',b',d'} \delta_2\left(q_{k,l}, A_{k,l}\left(i_{k,l}\right), q_{m,n}, A_{m,n}\left(i_{m,n}\right), p, b, d, p', b', d'\right)$$
$$\times \left|p, B, i_{k,l} + d\right\rangle \left\langle p', B', i_{m,n} + d'\right|$$

so that the RHS is a state. For any quantum configuration $\rho = \sum_k \lambda_k \rho_k$, Λ_{δ_2} is affine;

$$\Lambda_{\delta_2}\left(\sum_k \lambda_k \rho_k\right) = \sum_k \lambda_k \Lambda_{\delta_2}\left(\rho_k\right)$$

Definition 3. A GQTM M_{gq} is called unitary QTM (UQTM), if the quantum transition channel Λ_{δ_3} is a unitary channel implemented: $\Lambda_{\delta_3} \cdot = U_{\delta_3} \cdot U_{\delta_3}^*$, where U_{δ_3} is given by, for any $|\psi\rangle = |q, A, i\rangle$,

$$U_{\delta_3} |\psi\rangle = U_{\delta_3} |q, A, i\rangle$$
$$= \sum_{p,b,d} \delta_3\left(q, A\left(i\right), p, b, d\right) |p, B, i + d\rangle$$

where

$$\delta_3 : Q \times \Sigma \times Q \times \Sigma \times \{0, 1\} \to \mathbb{C}$$

is a transition function and it satisfies for any $q \in Q, a \in \Sigma, q'\,(\neq q) \in Q, a'\,(\neq a) \in \Sigma$,

$$\sum_{p,b,d} |\delta_3\left(q, a, p, b, d\right)|^2 = 1.$$

$$\sum_{p,b,d,d'} \delta_3\left(q', a', p, b, d'\right)^* \delta_3\left(q, a, p, b, d\right) = 0.$$

Remark 4. Classical Turing Machine (CTM) is represented as LQTM that the transition channel has only diagonal part. Moreover, for any $q, p \in Q, a, b \in \Sigma, d \in \{0, \pm 1\}$, put $\delta_3(q, a, p, b, d) = 0$ or 1 then UQTM is a reversal CTM.

The computational configuration is described by the density operator ρ on the Hilbert space. Applying the transition channel Λ_δ to the initial state, GTM proceeds the computation. When GTM goes to the final state such as accept or reject state, it halts with some probability. Essentially, the state change of configuration depends its last state, this channel Λ_δ has a state adaptivity.

2.2. *language classes of CTM*

Let L be a set of sequences of alphabet, we say L is *recognized* by a CTM M iff M halts for all $x \in L$ and does not halt for all $x \notin L$.

- P : recognized by a deterministic CTM in a polynomial time.
- NP : verifiable in polynomial time by a CTM.
- NP-complete : Most difficult problems in NP.
- NP-hard : if and only if there exists the NP-complete problem which is reduced to the class (e.g., general search problem).
- EXPTIME : a set of languages recognized by a deterministic CTM in an exponential time.

It holds the following inclusion relations.

$$P \subseteq NP \leq \text{NP-hard} \subset \text{EXPTIME}$$

where $A \leq B$ means that all languages in A is reduced to B.

2.3. *language classes of GTM*

Let us define the *recognition* of GQTM and some classes of languages.

Definition 5. Given a GQTM M_{gq} and a language L, if there exists N steps when M_{gq} recognizes L by the probability p, we say that the GQTM M_{gq} recognizes L by the probability p and its computational complexity is N.

Definition 6. A language L is bounded quantum probability polynomial time GQTM (BGQPP) if there is a polynomial time GQTM M_{gq} which accepts L with probability $p \geq \frac{1}{2}$.

Similarly, we can define the class of languages BUQPP (=BQPP) and BLQPP corresponding to UQTM and LQTM, respectively.

In the paper[10,14], it is shown that LQTM includes classical Turing machine, which implies

$$BPP \subseteq BLQPPL \subseteq BGQPP.$$

Moreover, if NLQTM accepts the SAT OV algorithm in polynomial time with probability $p \geq \frac{1}{2}$, then we have the inclusion

$$NP \subseteq BGQPP$$

3. Quantum Algorithm Using AD

A quantum algorithm is constructed by the following steps:

Step1. Prepare a Hilbert space $\mathcal{H} = \left(\mathbb{C}^2\right)^{\otimes n}$.
Step2. Construct an initial state $|\psi_{in}\rangle \in \mathcal{H}$.
Step3. Construct unitary operators U to solve the problem.
Step4. Apply them for the initial state and obtain a result state $|\psi_{out}\rangle = U |\psi_{in}\rangle$.
Step5. If necessary, amplify the probability of correct result.
Step6. Measure an observable with the result state.

The step 5 is important to obtain correct result when the success probability is very small. There are some methods to increase the probability.

3.1. *Chaos Amplifier*

Let $\rho = (1-p) |0\rangle \langle 0| + p |1\rangle \langle 1|$ be a state after measurement of quantum algorithm. Ohya and Volovich defined the Chaos Amplifier Λ_{CA} which is a non-linear classical process given by

$$\Lambda_{CA}(\rho) = \frac{(I + g_a(p)\sigma_3)}{2}$$

where σ_3 is the z-component of Pauli matrices, and

$$g_a(x) = ax(1-x)$$

It is proven in [2,3] that there exists a proper positive integer k such that

$$\text{tr}\left(\Lambda_{CA}\right)^k(\rho)\sigma_3 \geq \frac{1}{2}$$

The Chaos Amplifier is an example of state adaptive. Even if the probability p of correct result is very small ($p = \frac{1}{2^n}$), it can amplify very quick.

Theorem 7. *For a = 3.71, there exists an integer $k > \frac{5}{4}(n-1)$ such that*

$$tr\left(\Lambda_{CA}\right)^k (\rho)\,\sigma_3 \geq \frac{1}{2}$$

We proved that the Chaos Amplifier is represented by GKSL master equation in two qubit system[15].

4. Quantum Algorithm for Searching Problem

Let n be a positive integer, and we consider a searching problem with 2^n elements. Essentially, a searching problem belongs to NP-hard if we do not assume the existence of solution. We obtained the theorem[12].

The search problem in general case is defined as

Problem 8 *For a given f and $y \in Y$, we ask whether there exists $x \in X$ such that $f(x) = y$.*

Without loss of generality for discrete cases, we take $X = \{0, 1, \cdots, 2^n - 1\}$ and $Y = \{0, 1\}$. Here we use a discrete function f. Let n be a positive number, and f a function from $X = \{0, 1, \cdots, 2^n - 1\}$ to $Y = \{0, 1\}$.

We show a quantum algorithm to solve the problem. Denote x by the following binary expression

$$x = \sum_{k=1}^{n} 2^{k-1}\varepsilon_k, \tag{1}$$

where $\varepsilon_1, \cdots, \varepsilon_n \in \{0, 1\}$.

We divide the problem into several problems as below. Here we start the following problem:

Problem 9 *Does there exist x such that $f(x) = 1$ with $\varepsilon_1 = 0$?*

We go ahead to the next problem with the result of the above problem:

Problem 10 *Does there exist x such that $f(x) = 1$ with $\varepsilon_2 = 0$ for the obtained ε_1?*

After solving this problem, we know the value of ε_2, for example, when $\varepsilon_2 = 0$, x is written by $00\varepsilon_3 \cdots \varepsilon_n$ or $10\varepsilon_3 \cdots \varepsilon_n$.

Furthermore, we check the ε_i, $i = 3, \cdots, n$ by the same way as above using the information of the bits from ε_1 to ε_{i-1}. We run the algorithm from ε_1 to ε_n, and we look for one x satisfying $f(x) = 1$. Finally in the case that the result of the algorithm is $x = 1 \cdots 1$, we calculate $f(1 \cdots 1)$ and check whether $f(1 \cdots 1) = 1$ or not. We conclude that (1) if it becomes 1, $x = 1 \cdots 1$ is a solution of search problem, and (2) otherwise, there does not exist x such that $f(x) = 1$.

4.1. quantum algorithm for binary search

Let m be a positive integer which can be written by a polynomial in n. Let $\mathcal{H} = \left(\mathbb{C}^2\right)^{\otimes n+m+1}$ be a Hilbert space. We construct the following quantum algorithm $M_Q^{(1)}$ to solve the problem. Let $\left|\psi_{in}^{(1)}\right\rangle = |0^n\rangle \otimes |0^m\rangle \otimes |0\rangle \in \mathcal{H}$ be an initial vector for $M_Q^{(1)}$. The last qubit of $\left|\psi_{in}^{(1)}\right\rangle$ is for the answer, namely "yes" or "no". If the answer is "yes", then the last qubit becomes $|1\rangle$, otherwise $|0\rangle$. We start $M_Q^{(1)}$ with $\varepsilon_1 - 0$.

Step1. Apply Hadamard gates from the 2nd qubit to the n-th qubit.

$$I \otimes U_H^{\otimes n-1} \otimes I^{m+1} \left|\psi_{in}^{(1)}\right\rangle = \frac{1}{\sqrt{2^{n-1}}} |\varepsilon_1 (= 0)\rangle \otimes \left(\sum_{i=0}^{2^{n-1}-1} |e_i\rangle\right) \otimes |0^m\rangle \otimes |0\rangle$$

$$= \left|\psi_1^{(1)}\right\rangle, \tag{2}$$

where $|e_i\rangle$ are

$$|e_0\rangle = |0 \cdots 0\rangle$$
$$|e_1\rangle = |1 \cdots 0\rangle$$
$$\vdots$$
$$|e_{2^{n-1}-1}\rangle = |1 \cdots 1\rangle. \tag{3}$$

Step2. Let U_f be the unitary operator on $\mathcal{H} = \left(\mathbb{C}^2\right)^{\otimes n+m+1}$ to compute f, defined by

$$U_f |x\rangle \otimes |0^m\rangle \otimes |0\rangle = |x\rangle \otimes |z_x\rangle \otimes |f(x)\rangle, \tag{4}$$

where z_x is the dust qubit produced by the computation. Apply the unitary operator U_f to the state. The state becomes

$$U_f \left| \psi_1^{(1)} \right\rangle = \frac{1}{\sqrt{2^{n-1}}} |0\rangle \otimes \left(\sum_{i=0}^{2^{n-1}-1} |e_i\rangle \otimes |z_i\rangle \otimes |f(0e_i)\rangle \right)$$

$$= \left| \psi_2^{(1)} \right\rangle, \tag{5}$$

where z_i is the dust qubits depending on e_i.

Step3. Take the last qubit by unread measurement from the final state $\left| \psi_2^{(1)} \right\rangle$ such that

$$(1-p)|0\rangle\langle0| + p|1\rangle\langle1|, \tag{6}$$

where $p = card\{x|f(x) = 1, x = 0\varepsilon_2 \cdots \varepsilon_n\} / 2^{n-1}$

Step4. Apply the Chaos Amplifier to check whether the last qubit is in the state $|1\rangle\langle1|$.

After this algorithm, we know that if $\varepsilon_1 = 0$ or 1, then the last qubit is 1 or 0, respectively. We put this process as $M_Q^{(1)}(0^n) = \varepsilon_1$ where 0^n means the initial vector.

In generally, we write the algorithm $M_Q^{(i)}(\varepsilon_1, \varepsilon_2, \cdots, \varepsilon_{i-1}, 0^{n-i+1})$ for an initial vector $\left| \psi_{in}^{(i)} \right\rangle = |\varepsilon_1, \varepsilon_2, \cdots, \varepsilon_{i-1}, 0^{n-i+1}\rangle \otimes |0^m\rangle \otimes |0\rangle$ as the following:

Step1. Apply Hadamard gates from $i+1$-th to n-th qubits.

$$I^{\otimes i} \otimes U_H^{\otimes n-i} \otimes I^{m+1} \left| \psi_{in}^{(i)} \right\rangle = \frac{1}{\sqrt{2^{n-i}}} |\varepsilon_1, \varepsilon_2, \cdots, \varepsilon_{i-1}\rangle \otimes \left(\sum_{k=0}^{2^{n-i}-1} |e_k\rangle \right) \otimes |0^m\rangle \otimes |0\rangle$$

$$= \left| \psi_1^{(i)} \right\rangle. \tag{7}$$

Step2. Apply the unitary gate to compute f for the superposition made in the Step1, and store the result in $n+m+1$-th qubit.

$$U_f \left| \psi_1^{(i)} \right\rangle = \frac{1}{\sqrt{2^{n-i}}} |\varepsilon_1, \varepsilon_2, \cdots, \varepsilon_{i-1}\rangle \otimes \left(\sum_{k=0}^{2^{n-1}-1} |e_k\rangle \otimes |z_k\rangle \otimes |f(\varepsilon_1, \varepsilon_2, \cdots, \varepsilon_{i-1}, e_k)\rangle \right)$$

$$= \left| \psi_2^{(i)} \right\rangle. \tag{8}$$

Step3. Take the last qubit by the projection from the final state $\left| \psi_2^{(i)} \right\rangle$ such that

$$(1 - p) |0\rangle \langle 0| + p |1\rangle \langle 1| . \tag{9}$$

Step4. Apply the Chaos Amplifier to the amplitude p, so that we can easily find that the last qubit is $|1\rangle \langle 1|$.

After this algorithm $M_Q^{(i)} \left(\varepsilon_1, \varepsilon_2, \cdots, \varepsilon_{i-1}, 0^{n-i+1} \right)$, we know the bit ε_i such that $f(x) = 1$. Each $M_Q^{(i)}$, $i \geq 2$ use the result of all $M_Q^{(j)}$ $(j < i)$ as an initial vector. We run this algorithm $M_Q^{(i)}$ for each i $(i = 1, \cdots, n)$.

The computational complexity of this combined quantum algorithm is estimated as follows[12].

Theorem 11. *The number of unitary gates T in the quantum algorithm for searching problem becomes*

$$T = \frac{13}{8}n^2 - \frac{9}{4}n + nT(U_f)$$

where $T(U_f)$ is a number of unitary gates for the calculation of given objective function f.

References

1. M.Ohya and I.V.Volovich, J. opt. B, **5**,No.6 639-642, 2003.
2. M.Ohya and I.V.Volovich, Rep.Math.Phys., **52**, No.1,25-33 2003.
3. M.Ohya and I.V.Volovich, Mathematical Foundation of Quantum Computers, Teleportations and Cryptography, Springer, 2011
4. M.Ohya, QP-PQ:Quantum Prob. White Noise Anal. 21, Quantum Bio-Informatics, World Sci. Pub., 181-216, 2008
5. M.Asano, M.Ohya, Y.Tanaka, A.Khrennikov, I.Basieva, QP-PQ 28, Quantum Bio-Informatics IV, 51-60, 2011
6. M. Asano, I. Basieva, A. Khrennikov, M. Ohya, Y. Tanaka, and I. Yamato, Proceedings of Foundations of Probability and Physics-6, AIP 1244, 507-511, 2012
7. M. Asano, I. Basieva, A. Khrennikov, M. Ohya, Y. Tanaka, and I. Yamato, Quantum Interaction 2012, Lecture Notes in Computer Science Springer, 60-67, 2012
8. L.Accardi, Quantum Bio-informatics V, World Sci. Pub., QP-PQ 30, 1-13, 2013
9. S.Iriyama, T.Miyadera and M.Ohya, Phys. Lett. A, 372, 5120-5122, 2008.
10. S.Iriyama, M.Ohya and I.V.Volovich, QP-PQ:Quantum Prob. White Noise Anal., Quantum Information and Computing, 19, World Sci. Publishing, 204-225, 2006
11. S.Iriyama and M.Ohya, Open Syst. Inf. Dyn., 15, 2, 173-187, 2008

12. S. Iriyama, M. Ohya, and I.V. Volovich, Open Syst. Inf. Dyn. 22, 1550019, 2015
13. T. Hara, K.Sato and M. Ohya BMC Bioinformatics 11:235, 2010
14. S.Iriyama and M.Ohya, Open Syst. Inf. Dyn. 15:4, 383-396, 2008.
15. K.Furusho, S.Iriyama, Open Syst. Inf. Dyn., 24:2, 1750008, 2017

Quantum Bio-Informatics VI
© 2020 World Scientific Publishing Co. Pte. Ltd.
pp. 73–82

ON SOME CONSTRUCTIVE METHODS IN THE STUDIES OF COMPOUND STATES AND QUANTUM CHANNELS

ANDRZEJ JAMIOŁKOWSKI

Institute of Physics, Nicolaus Copernicus University,
87–100 Toruń, Poland
E-mail: jam@fizyka.umk.pl

The paper is dedicated to Professor Masanori Ohya
on the occasion of his retirement.

The concept of compound states was introduced and discussed by M. Ohya more than three decades ago [1]. It was connected with questions on how information contained in the input system is transmitted to the output system (cf. also [2]). In this paper some effective methods for the analysis of compound states and in particular of the Schatten decompositions used in Ohya's approach are presented. The role of these methods in formulation of quantum communication theory and quantum channels is discussed as well. In our discussion we use the concepts of partially commuting and partially equal operators.

1. Preliminaries and motivations

Let us recall some facts connected with the representations and evolutions of quantum states. Usually one represents a density operator $\varrho \in \mathcal{S}(\mathcal{H})$, where $\mathcal{S}(\mathcal{H})$ denotes the set of all density operators on a Hilbert space $\mathcal{H}$, by the spectral decomposition as

$$\varrho = \sum_k \lambda_k P_k. \qquad (1)$$

Here λ_k denotes an eigenvalue of ϱ and P_k is the projection from $\mathcal{H}$ onto the eigenspace associated with λ_k. If every eigenvalue λ_k is nondegenerate, then the dimension of the range of P_k is one. If a certain eigenvalue, say λ_m, is degenerate, then P_m can be further decomposed into one-dimensional projections

$$P_m = \sum_{j=1}^{\kappa} E_j^{(m)}, \qquad (2)$$

where κ denotes the dimension of the range of P_m and $E_j^{(m)}$ is a one-dimensional projection given by

$$E_j^{(m)} = |e_j^{(m)}\rangle\langle e_j^{(m)}|. \qquad (3)$$

Here $|e_j^{(m)}\rangle$ for $j = 1, 2, \ldots, \kappa$ represent eigenvectors belonging to the range of P_m. By relabelling of indices we can write $\varrho = \sum_j \lambda_j E_j$, where $\lambda_1 \geq \lambda_2 \geq \ldots \geq \lambda_j \geq \ldots$, and $E_i \perp E_j$ $(i \neq j)$. This kind of decomposition is said to be the Schatten decomposition. In this description the eigenvalue of multiplicity k is repeated precisely k times.

Now, when a state ϱ changes to another state $\tilde{\varrho}$ one can ask how much information of ϱ is correctly transmitted to $\tilde{\varrho}$, i.e. we consider relations

$$\varrho \mapsto \tilde{\varrho} = \Phi(\varrho). \qquad (4)$$

For this purpose the notion of the compound state was introduced by M. Ohya in 1983 [1–3].

At this point some comments should be useful.

According to fundamental assumptions of information theory all dynamical effects for information transmission are represented by channels. In Shannon's theory, that is in the classical case, information of the system is carried by a probability distribution of events, and the channel introduces a change of this probability distribution. The concept of states in quantum information theory plays a role of an extension of that probability distribution. Therefore one assumes that information in the quantum case is carried by a state (density operator) and the channel provides a dynamical change of states.

Let us assume that an input state ϱ is transmitted through a channel Φ to an output system and the resulting state is $\Phi(\varrho)$. The state ϱ carries all information of the state of the input system and we would like to know how much information is correctly sent to the output system. For this purpose one has to analyse some relations between operators of ϱ and $\tilde{\varrho} = \Phi(\varrho)$. To this end the concept of the compound state was introduced. According to the definition introduced by Ohya in [1] the compound state is given by expression of the form

$$\varrho_E = \sum_j \lambda_j E_j \otimes \Phi(E_j) \tag{5}$$

which in an obvious way depends on the Schatten decomposition of ϱ.

In the algebraic formulation of quantum mechanics, a fixed quantum mechanical system is represented by an algebra $\mathcal{A}$ of operators acting on some Hilbert space $\mathcal{H}$. In this approach, the observables of the system are identified with self-adjoint elements in $\mathcal{A}$ and the physical states are given by positive unital functionals on $\mathcal{A}$. We will consider the case when $\mathcal{H}$ is finite-dimensional, then the set of states can be identified with the set of $\mathcal{S}(\mathcal{H})$ of density operators, that is, positive elements in $\mathcal{A}$ with unital trace. The evolutions of the system are described by transformations on the set $\mathcal{S}(\mathcal{H})$, or more generally, by linear maps between $\mathcal{S}(\mathcal{H})$ and $\mathcal{S}(\mathcal{K})$, where $\mathcal{H}$ and $\mathcal{K}$ represent two finite-dimensional Hilbert spaces. In this formulation the compound state η of the tensor product of states ϱ and $\tilde{\varrho}$ should satisfy the following properties:

(1) $\eta(A \otimes \mathrm{I}) = \varrho(A)$ for any $A \in \mathcal{A}$;
(2) $\eta(\mathrm{I} \otimes B) = \tilde{\varrho}(B)$ for any $B \in \tilde{\mathcal{A}}$;
(3) the expression of η contains the classical expression as a special case;

In 1. and 2. $\mathcal{A}$ and $\tilde{\mathcal{A}}$ denote algebras of observables. When $\mathcal{A}$ is the full algebra $B(\mathcal{H})$ any state ϱ gives us corresponding expectation values $\varrho(A) = \mathrm{Tr}(\varrho A)$ for any $A \in \mathcal{A}$.

The Schatten decomposition of $\varrho = \sum_j \lambda_j E_j$, $E_j = |x_j\rangle\langle x_j|$, is not unique unless every eigenvalue λ_j is nondegenerate. The Ohya's compound state η is given by

$$\eta_E(Q) = \mathrm{Tr}\,\varrho_E Q, \quad Q \in \mathcal{A} \otimes \tilde{\mathcal{A}} \tag{6}$$

and, indeed, we obtain

$$\varrho_E := \sum_j \lambda_j E_j \otimes \Phi(E_j), \tag{7}$$

where E represents a Schatten decomposition $\{E_j\}$. In this case, the mutual entropy for ϱ and the channel Φ is expressed by

$$I(\varrho, \Phi) = \sup \{I_E(\varrho, \Phi); \quad E = \{E_j\}\} \tag{8}$$

with

$$I_E(\varrho, \Phi) = \mathrm{Tr}\,\varrho_E (\ln \varrho_E - \ln \varrho_0). \tag{9}$$

2. Effective procedures

Let $A \in M_n(\mathbb{C})$. It is well known that if $n \geq 5$ then the problem of calculation of eigenvalues and eigenvectors of A is *in general* not solvable in an effective way, that is, by a finite number of arithmetic operations on the elements of A. This observation restricts to some extent usefulness of the Schatten decompositions.

However, a lot of spectral properties of the matrix (operator) A can be analyzed by some effective procedures.

At this point we would like to add some remarks. In many problems of physics (and, in particular, in quantum information theory) one defines some objects or describes their properties by infinite number of conditions. A typical example is the condition of positive semidefiniteness of a quantum state ϱ which, even in a finite-dimensional Hilbert space $\mathcal{H}$, one defines by the infinite number of inequalities, namely $\langle x, \varrho x \rangle \geq 0$ for all $|x\rangle \in \mathcal{H}$, where $\langle , \rangle$ denotes an inner product in $\mathcal{H}$. Fortunately, in many situations the infinite number of conditions can be replaced by checking a finite number of equalities or inequalities. In such cases we say that we are able to introduce some effective procedures (methods). In case of positivity of matrices these are the Sylvester criteria.

In other words, by an "effective method" we mean such a procedure that can be done without any approximations; it can be long, it can be complicated, but it can be realized by a *finite* number of arithmetic calculations. Our aim is to show that such effective procedures are possible in many situations which we meet in quantum information theory.

As an example, let us take an arbitrary $A \in M_n(\mathbb{C})$. It is obvious that the characteristic polynomial and the minimal polynomial of A can be calculated by effective procedures. In this way, we can find answers to some questions connected with the spectral structure of A, where these polynomials are important. We recall that according to the so-called Abel-Ruffini theorem there is no general algebraic solution to polynomial equations of degree five or higher with arbitrary coefficients [4]. However, this theorem does not state that some higher-order polynomial equations have no solution. The Abel-Ruffini theorem only states that there is no general method (general procedure) that applies to all equations of degree five or higher.

Now, one can formulate for $A \in M_n(\mathbb{C})$ the following problems.

(1) Are all eigenvalues of A distinct or not?
(2) Can we diagonalize A or not?
(3) Is the matrix A derogatory or nonderogatory?

Let us recall that matrix is called nonderogatory if each eigenvalue of $A \in M_n(\mathbb{C})$ has geometric multiplicity exactly 1 (regardless of the algebraic multiplicity).

The geometric multiplicity is just the maximum number of linearly independent eigenvectors associated with an eigenvalue.

Now, we meet similar problems if we want to calculate common eigenvalues or eigenvectors of two matrices A and B (for example ϱ and $\tilde{\varrho} = \Phi(\varrho)$). However, if we want to establish only the fact of existence of common eigenvalues for ϱ and $\tilde{\varrho}$ without knowing them, the solution to this problem can be obtained by a simple effective procedure.

The basic tools in the detection of existence of some degenerate eigenvalues of ϱ or some common invariant subspaces of ϱ and $\tilde{\varrho}$ are discriminants of ϱ and the so-called Shemesh criterion, respectively (cf. e.g. [7]).

3. The discriminants and resultants

The Schatten decomposition of ϱ can be analyzed by using the notion of the discriminant of a matrix. In the beginning, in the nineteenth century, the concept of discriminant was applied to polynomials in one variable.

Let π denote a polynomial and $\{\lambda_i\}$ are the zeros of π. Then

$$\mathrm{Disc}(\pi) := \prod_{i<j}(\lambda_i - \lambda_j)^2 . \tag{10}$$

The importance of $\mathrm{Disc}(\pi)$ is that it appears to be a polynomial in π's coefficients whereas the zeros $\{\lambda_i\}$ are, in general, algebraic functions of those coefficients.

The discriminant of a square matrix $A \in M_n(\mathbb{C})$, $\mathrm{Disc}(A)$, is defined as the discriminant of its characteristic polynomial $\chi(A)$. However, one can avoid the use of $\chi(A)$. We will define a matrix with the entries given directly from the entries of A, whose determinant yields $\mathrm{Disc}(A)$.

Let $\varrho \in S(\mathcal{H})$, where $S(\mathcal{H})$ be defined by

$$S(\mathcal{H}) := \{\varrho : \mathcal{H} \rightarrow \mathcal{H}, \quad \varrho \geq 0, \quad \mathrm{Tr}\, \varrho = 1\}. \tag{11}$$

If ϱ is a fixed density matrix with the eigenvalues $\lambda_1, \ldots, \lambda_n$, then

$$\mathrm{Disc}(\varrho) := \prod_{i<j}(\lambda_i - \lambda_j)^2 . \tag{12}$$

We look for description of $\mathrm{Disc}(\varrho)$ in terms of the entries of ρ.

Let us consider a nonlinear transformation on $S(\mathcal{H})$ defined by $\varrho \rightarrow R$, where R denotes the matrix with entries

$$R_{ij} := \mathrm{Tr}\left(\varrho^{i+j-2}\right). \tag{13}$$

Then one can show (B. N. Parlett 2002 and N. V. Ilyushechkin 1992) that

$$\text{Disc}(\varrho) = \det R. \tag{14}$$

Moreover, we can express all R_{ij} as special multiplications

$$
\begin{aligned}
R_{ij} &= \text{Tr}\left[\varrho^{i-1} \cdot \varrho^{j-1}\right] \\
&= \text{vec}\left[(\varrho^T)^{i-1}\right] \cdot \text{vec}[\varrho^{j-1}].
\end{aligned} \tag{15}
$$

If instead of analyzing the matrix ϱ we want to compare the eigenvalues of ϱ and $\tilde{\varrho}$, or more generally, the eigenvalues of matrices A and B from $M_n(\mathbb{C})$, then the fact of existence or absence of common eigenvalues — without calculating them — can be checked by an effective method. In this case the procedure consists in finding characteristic polynomials $\chi(\varrho)$ and $\chi(\tilde{\varrho})$ (or more generally $\chi(A)$ or $\chi(B)$) and their resultant (that is, the determinant of Sylvester matrix constructed from the characteristic polynomials of A and B). Details of these procedures are described in [8, 9] and [10].

4. Partially commuting and partially equal operators

As earlier, let ϱ and $\tilde{\varrho}$ denote the input and output states (or A and B any two matrices), respectively. One can ask if there exist proper subspaces in $\mathbb{C}^n$ which are invariant with respect to A and B and such that both matrices (states) are identical if they are restricted to these subspaces. Of course, in this case the matrices A and B must commute on the considered subspaces — they are partially commuting. In other words, we ask about the proper subspaces $\mathcal{N}$ of $\mathbb{C}^n$, i.e. such that $\mathcal{N} \subsetneq \mathbb{C}^n$,

$$1)\ A(\mathcal{N}) \subset \mathcal{N}, \quad 2)\ B(\mathcal{N}) \subset \mathcal{N}$$

and at the same time $ABx = BAx$ for all $x \in \mathcal{N}$. If in addition

$$A|_{\mathcal{N}} = B|_{\mathcal{N}}, \tag{16}$$

then we will call such operators the partially identical operators and a subspace satisfying 1) and 2) together with (16) the subspace on which the operators A and B agree (they are equal).

We have an example of such a subspace if the conditions

$$Ax = \lambda x \quad \text{and} \quad Bx = \lambda x \tag{17}$$

are satisfied. In this case $\mathcal{N}$ is a minimal A-, B-invariant subspace which contains x and on which A and B commute. As is well known the maximal subspace $\mathcal{M}$ on which operators A and B commute is given by the so-called Shemesh criterion, namely [11]

$$\mathcal{M} := \bigcap_{k,l=1}^{n-1} \mathrm{Ker}\left(\left[A^k, B^l\right]\right), \tag{18}$$

where n, in the intersection, can be replaced by p and q, the degrees of the minimal polynomials of A and B and $[A, B]$ denotes the commutator of A and B.

Let us observe that we can construct the matrix

$$\Omega := \sum_{k,l=1}^{n-1} \left[A^k, B^l\right]^* \left[A^k, B^l\right], \tag{19}$$

for which we have the relation $\mathcal{M} = \mathrm{Ker}\,\Omega$. In this way one obtains an effective procedure for construction of $\mathcal{M}$.

The maximal subspace $\mathcal{N}$ satisfying conditions 1), 2) and (16) we will denote $C(A, B)$ because the subspace is the maximal one on which the operators coincide.

On the base of paper by D. Shemesh one can show that

$$C(A, B) = \left(\bigcap_{k=1}^{p} \bigcap_{l=1}^{q} \mathrm{Ker}\left(\left[A^k, B^l\right]\right)\right) \bigcap \mathrm{Ker}(A - B), \tag{20}$$

where p denotes the degree of the minimal polynomial of A, q denotes the degree of the minimal polynomial of B.

The formula (20) is constructed on the base of the following argumentation:

1) As was said above according to [11] the subspace

$$\mathcal{M} := \bigcap_{k=1}^{p} \bigcap_{l=1}^{q} \mathrm{Ker}\left(\left[A^k, B^l\right]\right) \tag{21}$$

is the maximal subspace of $\mathbb{C}^n$ which is invariant with respect to A and B and on which the operators A and B commute.

2) In our problem we are interested in the case when operators A and B not only commute on $\mathcal{M}$ but even are identical. In this case we look for the subspace of $\mathbb{C}^n$ on which the operators coincide.

It is not difficult to see that condition (20) can be restated in the following constructive form. The subspace $C(A, B)$ is the kernel of the matrix

$\tilde{\Omega}$ which we can obtain by observation that $C(A, B) = \mathcal{N}$, where

$$\mathcal{N} := \bigcap_{k=1}^{n} \mathrm{Ker}\left(\left[A^k - B^k\right]\right) . \tag{22}$$

Indeed, the subspace $\mathcal{N}$ of the above form one can obtain by applying the Shemesh criterion (18) to two matrices

$$K := \begin{bmatrix} A & 0 \\ 0 & B \end{bmatrix} \quad \text{and} \quad J := \begin{bmatrix} 0 & I \\ I & 0 \end{bmatrix} . \tag{23}$$

In this case, K^r is given by powers A^r and B^r and we have $J^{2m} = I$ for all $m = 1, 2, \dots$. Here, I denotes the identity in $2n$-dimensional space. On the other hand $\tilde{\Omega}$, in analogy to (19), is constructed by

$$\tilde{\Omega} := \sum_{k=1}^{n} \left[A^k - B^k\right]^* \left[A^k - B^k\right] \tag{24}$$

which means that the subspace $\mathcal{N}$ can be constructed in an effective way.

5. Compound operators

In this section we will discuss the problem of existence of common invariant subspaces for more than two operators.

In order to discuss common invariant subspaces of operators of dimension 2 or more for a set of operators $A_1, \dots, A_s$ we use the concept of compound operators.

In the space $\bigotimes^k \mathcal{H}$ there is an especially important subspace $\bigwedge^k \mathcal{H}$ (kth-Grassmann subspace). The antisymmetric tensor product of vectors $x_1, \dots, x_k$ in $\mathcal{H}$ is defined as

$$x_1 \wedge \dots \wedge x_k := (k!)^{-1/2} \sum_{\sigma} \epsilon_{\sigma} x_{\sigma(1)} \otimes \dots \otimes x_{\sigma(k)}, \tag{25}$$

where σ runs over all permutations of the k indices and ϵ_{σ} is ± 1, depending on whether σ is an even or odd permutation.

The span of all antisymmetric tensors $x_1 \wedge \dots \wedge x_k$ in $\bigotimes^k \mathcal{H}$ is denoted by $\bigwedge^k \mathcal{H}$ and is called kth exterior product of the space $\mathcal{H}$.

One can see that the subspace $\bigwedge^k \mathcal{H}$ is invariant under the operator $\bigotimes^k A$,

$$\overset{k}{\bigotimes} A : \overset{k}{\bigwedge} \mathcal{H} \to \overset{k}{\bigwedge} \mathcal{H}. \tag{26}$$

The restriction of $\bigotimes^k A$ to this invariant subspace is denoted by $C_k(A)$, and is called kth compound operator of A (or kth Grassmann power of A). On elementary antisymmetric tensors $x_1 \wedge \ldots \wedge x_k$ we have

$$C_k(A)(x_1 \wedge \ldots \wedge x_k) = Ax_1 \wedge \ldots \wedge Ax_k. \tag{27}$$

Simple properties of these operators can be summarized as:

$$C_k(AB) = C_k(A)C_k(B), \quad C_k(A^*) = C_k(A)^*, \quad C_k(A^{-1}) = C_k(A)^{-1}. \tag{28}$$

If A is hermitian, unitary, normal or positive, then so are $C_k(A)$.

A vector $x \in \bigwedge^k \mathcal{H}$ is called decomposable iff $x = x_1 \wedge \ldots \wedge x_k$ for some $x_i \in \mathcal{H}$, $i = 1, \ldots, k$; one refers to $x_1, \ldots, x_k$ as factors of x.

By properties of $C_k(A)$, those decomposable vectors whose factors are linearly independent eigenvectors of A are eigenvectors of $C_k(A)$.

The spectrum of $C_k(A)$ coincides with the set of all possible k-products of the eigenvalues of A.

Theorem 5.1. *[generalization of the Shemesh criterion]*
Let A_1 and A_2 belong to $\mathcal{B}(\mathcal{H})$ and let them be invertible. Then A_1 and A_2 have a common invariant subspace of dimension k, for $k \geq 2$, iff $C_k(A_1)$ and $C_k(A_2)$ have a common decomposable eigenvector.

Theorem 5.2. *Assume that $A_1, \ldots, A_s \in \mathcal{M}_n(\mathbb{C})$ and*

$$\mathcal{M}(A_1, \ldots, A_s) := \bigcap_{\substack{k_i, l_j \geq 0 \\ k_1+k_2+\ldots+k_s \neq 0 \\ l_1+l_2+\ldots+l_s \neq 0}}^{n-1} \mathrm{Ker}\left[A_1^{k_1} \ldots A_s^{k_s}, A_1^{l_1} \ldots A_s^{l_s}\right]. \tag{29}$$

(1) Matrices A_i have a common eigenvector if and only if

$$\mathcal{M}(A_1, \ldots, A_s) \neq 0. \tag{30}$$

(2) We have $\mathcal{M}(A_1, \ldots, A_s) = \mathrm{Ker}\,\Omega$, where

$$\Omega = \sum_{\substack{k_i, l_j \geq 0 \\ k_1+k_2+\ldots+k_s \neq 0 \\ l_1+l_2+\ldots+l_s \neq 0}}^{n-1} [A_1^{k_1} \ldots A_s^{k_s}, A_1^{l_1} \ldots A_s^{l_s}]^*[A_1^{k_1} \ldots A_s^{k_s}, A_1^{l_1} \ldots A_s^{l_s}]. \tag{31}$$

If $d > 1$ and each of $A_1, \ldots, A_s \in \mathcal{M}_n(\mathbb{C})$ has pairwise different eigenvalues, we can apply Theorem 5.2 to establish a computable criterion for the existence of common invariant subspace of A_i of dimension d.

Details connected with Theorem 5.1 one can find in the paper by M. Tsatsomeros [12] and Theorem 5.2 is formulated and proven in A. Jamiołkowski and G. Pastuszak [13, 14].

6. Comments

The main point of this paper is discussed in Section 4, where the concept of partially equal (identical) operators is introduced. This idea is not only important from theoretical point of view but has also some natural applications in analysing properties of quantum channels [15]. Moreover, there exists a constructive way to check the existence of subspaces on which two operators coincide. One can also extend this method to a number of operators greater than two. The notions of partially commutative and partially equal subspaces give new possibilities in the discussion of some problems in communication theory. The Ohya papers [1, 2] opened these approaches.

References

1. M. Ohya: *IEEE Trans. Information Theory* **29** (1983) 19.
2. M. Ohya: *Rep. Math. Phys.* **27** (1989) 19.
3. M. Ohya, I. Volovich: *Mathematical Foundations of Quantum Information and Computation*, Springer 2011.
4. N. Jacobson: *Basic Algebra* 1, Dover 2009 (2nd ed.).
5. B. N. Parlett: *Lin. Algebra and its Appl.* **355** (2004) 85.
6. N. V. Ilyushechkin: *Math. Notes* **51** (1992) 230.
7. Kh. D. Ikramov: *J. of Math. Sciences* **150** (2008) 1937.
8. S. Basu et al.: *Algorithms in Real Algebraic Greometry*, Springer 2006.
9. I. M. Gelfand et al.: *Discriminants, Resultants, and Multidimensional Determinants*, Birkhäuser 1994.
10. D. Cox et al.: *Ideals, Varieties, and Algorithms*, Springer 2007.
11. D. Shemesh: *Lin. Algebra and its Appl.* **62** (1984) 11.
12. M. Tsatsomeros: *Lin. Algebra and its Appl.* **322** (2001) 51.
13. A. Jamiołkowski, G. Pastuszak: *Lin. Algebra and its Appl.* **63** (2015) 314.
14. G. Pastuszak, A. Jamiołkowski: *Electron. J. Lin. Algebra* **30** (2015) 253.
15. A. Jamiołkowski: in preparation.

Quantum Bio-Informatics VI
© 2020 World Scientific Publishing Co. Pte. Ltd.
pp. 83–86

INFORMATION DYNAMICS AND ENTROPIC CHAOS DEGREE FOR THE ANALYSIS OF DYNAMICAL SYSTEMS

TAKEO KAMIZAWA

Department of Information Sciences, Tokyo University of Science,
Noda City Chiba 278-8510, Japan
E-mail: kamizawa@rs.tus.ac.jp

Information dynamics (ID) proposed by Ohya opened a broader view towards complex systems[1-5], where ID provides a general framework of how we see such complicated systems and some quantities representing 'complexities'. In each field of science there are quantities describing certain 'complexities' (e.g. entropy, computational complexity, etc.), but they are usually model-dependent and often they are not comparable to each other. Since ID is a general scheme, it enables us to treat these independent 'complexities' in the same framework.

Let $(\mathcal{A}, \mathfrak{S}(\mathcal{A}), \alpha(G))$ be an input system (i.e. $\mathcal{A}$ is some set of 'objects' to be observed, $\mathfrak{S}$ is a set of methods to obtain the observed values and $\alpha(G)$ describes a certain evolution of the system with a set G of parameters), and suppose the output system is identical to the input system, for simplicity. Let $\mathcal{S} \subset \mathfrak{S}(\mathcal{A})$ (this is called a reference system) and $\Lambda^* : \mathfrak{S} \to \mathfrak{S}$ be some map. Fundamental quantities in ID are the (i) 'complexity of a state' $\varphi \in \mathcal{S}$ and (ii) 'transmitted complexity' of $\varphi \mapsto \Lambda^*(\varphi)$. Usually these complexities are denoted by $C^{\mathcal{S}}$ and $T^{\mathcal{S}}$, respectively, and both of them are required to satisfy certain axioms[5]: For any $\varphi \in \mathcal{S} \subset \mathfrak{S}$,

(1) $C^{\mathcal{S}}(\varphi) \geq 0$, $T^{\mathcal{S}}(\varphi; \Lambda^*) \geq 0$.
(2) $0 \leq T^{\mathcal{S}}(\varphi; \Lambda^*) \leq C^{\mathcal{S}}(\varphi)$.
(3) $T^{\mathcal{S}}(\varphi; \mathrm{id}) = C^{\mathcal{S}}(\varphi)$, ($\mathrm{id} : \mathfrak{S} \to \mathfrak{S}$ is the identity map).

(4) For any map $j : \mathfrak{S} \to \mathfrak{S}$ such that $j \mid_{\text{ext}\mathcal{S}}$ is a disjoint bijection,

$$C^{j(\mathcal{S})} \left(j \left(\varphi \right) \right) = C^{\mathcal{S}} \left(\varphi \right)$$
$$T^{j(\mathcal{S})} \left(j \left(\varphi \right) ; \Lambda^* \right) = T^{\mathcal{S}} \left(\varphi ; \Lambda^* \right).$$

(5) For $\Phi = \varphi \otimes \psi \in \mathcal{S}_{\text{tot}} \subset \mathfrak{S}_{\text{tot}}$ ($\mathfrak{S}_{\text{tot}}$ is the total system),

$$C^{\mathcal{S}_{\text{tot}}} \left(\Phi \right) = C^{\mathcal{S}} \left(\varphi \right) + C^{\mathcal{S}} \left(\psi \right),$$

where $\otimes$ is a symbolic expression for the combination of two states.

For the analysis of complex systems, ID provides an important quantity, namely the chaos degree (CD). This is a general concept for measuring the complexity of the state change and it naturally arises in the theory of ID. For a complex system, let $\mathcal{S} \subset \mathfrak{S}$ be a reference system and $\Lambda^* : \mathfrak{S} \to \mathfrak{S}$ be a map. For each $\varphi \in \mathcal{S}$, it is natural to say that the quantity:

$$D^{\mathcal{S}} \left(\varphi ; \Lambda^* \right) = C^{\mathcal{S}} \left(\Lambda^* \left(\varphi \right) \right) - T^{\mathcal{S}} \left(\varphi ; \Lambda^* \right)$$

describes the 'complexity increased during the evolution Λ^*'. Then, the quantity $D^{\mathcal{S}}$ is called the chaos degree (CD) with respect to φ. If $\mathfrak{S}$ possesses a topological structure and $\mathcal{S}$ is a compact convex set, each state $\varphi \in \mathcal{S}$ has an extremal decomposition (Choquet representation)

$$\varphi = \int_{\mathcal{S}} \omega d\mu,$$

where μ is a measure, and the set of all possible measures is denoted by $M_\varphi \left(\mathcal{S} \right)$. Then, the CD is given by

$$D^{\mathcal{S}} \left(\varphi ; \Lambda^* \right) = \inf \left\{ \int_{\mathcal{S}} C^{\mathcal{S}} \left(\Lambda^* \omega \right) d\mu \mid \mu \in M_\varphi \left(\mathcal{S} \right) \right\}.$$

As described above, ID and CD are general frameworks, so in order to apply to the analysis of some specific complex system, we need to mathematically define input/output systems $\left(\mathcal{A}, \mathfrak{S} \left(\mathcal{A} \right), \alpha \left(G \right) \right)$, the reference system $\mathcal{S}$, the dynamics Λ^* and the complexities $C^{\mathcal{S}}, T^{\mathcal{S}}$. An important example is constructed if we set $C^{\mathcal{S}} \left(\varphi \right) = S \left(\varphi \right)$ (Shannon entropy) and $T^{\mathcal{S}} \left(\varphi ; \Lambda^* \right) = I \left(\varphi ; \Lambda^* \right)$ (mutual entropy). In this case, one can check that the Shannon entropy and the mutual entropy satisfy the axioms above, and the CD of this special type:

$$D^{\mathcal{S}} \left(\varphi ; \Lambda^* \right) = S \left(\Lambda^* \left(\varphi \right) \right) - I \left(\varphi ; \Lambda^* \right)$$

is called the entropic chaos degree (ECD).

ECD can be used to analyse complicated behaviour in dynamical systems[2–8]. Let X be a phase space, $f : X \to X$ be a map and consider a trajectory $\mathcal{O}_{x_0} = (x_0, f(x_0), f^2(x_0), \ldots)$. By some countable partition $\tilde{A} = \{A_j\}_j$ of X, a 'state of the trajectory' may be described by a probability distribution with respect to $\tilde{A}$. Hence, if the frequency of points in the trajectory $\mathcal{O}_{x_0}$ being in $A_j \in \tilde{A}$ is denoted by p_j, the state complexity is

$$C(f\mathcal{O}_{x_0}) = S\left((p_j)_j\right) = -\sum_j p_j \log p_j.$$

Notice that $C(f\mathcal{O}_{x_0}) = C(\mathcal{O}_{x_0})$ if the length of the trajectory is infinite because p_j does not change. The 'transmitted complexity of the trajectory' is the mutual entropy, so if p_{jk} denotes the frequency of points in the trajectory being in $A_j, A_k \in \tilde{A}$ continuously in this order, we have

$$T(\mathcal{O}_{x_0}; f) = I\left((p_{jk})_{j,k}\right) = \sum_{j,k} p_{jk} \log \frac{p_{jk}}{p_j p_k}.$$

Therefore, the ECD of the dynamical system with respect to the trajectory is

$$D_{\tilde{A}}(x_0, f) \stackrel{\text{def}}{=} D^S = C(f\mathcal{O}_{x_0}) - T(\mathcal{O}_{x_0}; f)$$

$$= -\sum_k \left(\sum_j p_{jk}\left(\log p_k + \log \frac{p_{jk}}{p_j p_k}\right)\right)$$

$$= -\sum_{j,k} p_{jk} \log \frac{p_{jk}}{p_j}. \tag{1}$$

ECD can well detect complicated behaviour of trajectories in dynamical systems as many examples show[2–8].

More generally, the probability distribution $(p_{i_0 \cdots i_{N-1}})$, where $p_{i_0 \cdots i_{N-1}}$ denotes the frequency of points in the trajectory being in $A_{i_0}, \ldots, A_{i_{N-1}}$ continuously in this order for some N, can be used to represent the state of the trajectory $\mathcal{O}_{x_0}$. In this case, a similar discussion shows the ECD is:

$$D_{\tilde{A}}^N(x_0, f) = -\sum_{i_0, \ldots, i_{N-1}} p_{i_0 \cdots i_{N-1}} \log \frac{p_{i_0 \cdots i_{N-1}}}{p_{i_0 \cdots i_{N-2}}}$$

and this is called the ECD with N-step memory effect. According to some numerical experiments, the value of ECD had been conjectured to be equal to the value of the (largest) Lyapunov exponent (LE), which is another well-known criterion of complexity. Under some specific condition, relations among ECD, LE and another criterion so-called the Kolmogorov-Sinai entropy (KS) were proved[9]:

Theorem 1. *Suppose M is an n-dimensional smooth compact Riemannian manifold and $f : M \to M$ be a C^1-diffeomorphism with an ergodic invariant Borel probability measure μ. Then, ECD, KS, LE satisfy:*

$$\lim_{N \to \infty} D_{\tilde{A}}^N (x, f) \leq S_\mu (f) \leq \sum_{\lambda_k \geq 0} \lambda_k (x) \ (\mu\text{-}a.e.) ,$$

where $\tilde{A}$ is some measurable countable partition of M, $D_{\tilde{A}}^N (x, f)$ is the ECD with N-step memory effect, S_μ is the KS and $\lambda_1, \ldots, \lambda_n$ are the LE of the system.

The inequalities enable us to estimate the value of some criterion from another. ECD is an advanced tool from the computational point because the computations of the probability distributions and the equation (1) are quite simple compared to the computations of the other criteria. The inequality above can help to estimate the values of the KS and the LE using the ECD.

References

1. Ingarden, R. S., Kossakowski, A. and Ohya, M., *Information Dynamics and Open Systems*, Kluwer Academic Publishers, 1997.
2. Ohya, Masanori, Information dynamics and its applications to optical communication processes, Quantum Aspects of Optical Communications, pp.81–pp.92, Springer, 1991.
3. Ohya, Masanori, Complexities and their applications to characterization of chaos, International Journal of Theoretical Physics, Vol.37, No.1, pp.495–pp.505, Springer, 1998.
4. Ohya, M., Information Dynamics and Its Applications (White Noise Analysis and Quantum Probability), RIMS Kokyuroku, Vol.874, 1994.
5. Ohya, Masanori and Volovich, Igor., *Mathematical foundations of quantum information and computation and its applications to nano-and bio-systems*, Springer, 2011.
6. Inoue, Kei and Ohya, Masanori and Sato, Keiko, Application of chaos degree to some dynamical systems, Chaos, Solitons and Fractals, Vol.11, No.9, pp.1377–pp.1385, Elsevier, 2000.
7. Kossakowski, A. and Ohya, M. and Togawa, Y., How can we observe and describe chaos?, Open System and Information Dynamics, Vol.10, pp.221–pp.233, 2003.
8. Ohya, Masanori, Complexity in Dynamics and Computation, Acta Applicandae Mathematica, Vol.63, No.1, pp.293–pp.306, 2000.
9. Kamizawa, T. and Hara, T. and Ohya, M., On relations among the entropic chaos degree, the Kolmogorov-Sinai entropy and the Lyapunov exponent, Journal of Mathematical Physics, Vol.55, No.3, AIP, pp.032702, 2014.

Quantum Bio-Informatics VI
© 2020 World Scientific Publishing Co. Pte. Ltd.
pp. 87–92

HAS CHSH-INEQUALITY ANY RELATION TO EPR-ARGUMENT?

ANDREI KHRENNIKOV

*International Center for Mathematical Modeling
in Physics, Engineering, Economics, and Cognitive Science
Linnaeus University, Växjö, Sweden*

We emphasize the role of the precise correlations loophole in attempting to connect the CHSH-type inequalities with the EPR-argument. The possibility to test theories with hidden variables experimentally by using such inequalities is questioned. The role of the original Bell inequality is highlighted. The interpretation of the CHSH-inequality in the spirit of Bohr, as a new test of incompatibility, is presented. The positions of Bohr, Einstein, Podolsky, Rosen, Bell, Clauser, Horne, Shimony, Holt, and De Broglie are enlightened.

Keywords: EPR-argument; complementarity and nonlocality; original Bell inequality; CHSH-inequality; perfect correlations; hypothetical subquantum correlations.

1. Introduction

The recent success in performing clean and loophole free experiments[1,2,3] testing violations of the Bell-type inequalities can make the impression that the long debate on interpretation of violation of these inequalities has been finally ended. Moreover, some authors[4,5], consider these experiments as the final accords in the long debate between Einstein and Bohr. Such authors couple these inequalities with the Einstein-Podolsky-Rosen (EPR) framework[6], cf., however, with works[7,8]. Nowadays it is widely claimed that *"Bohr was right and Einstein was wrong"*. It is interesting that this formulation peacefully coexists with the statement that experiment confirms *"quantum nonlocality.* The aim of this note is analyzing the After-Bell situation in quantum foundations, see also article[8].

2. Does CHSH-inequality have any relation to the EPR framework?

I claim that the answer to this question is negative. The key point of the EPR framework is consideration of *precise correlations* and coupling them with EPR elements of reality. In particular, the EPR statement on quantum nonlocality (as an absurd alternative to incompleteness of QM) has meaning only this framework. Those few who read the original Bell's paper[9], see also book[10], know that here Bell tried to mimic the EPR framework[6] by using hidden variables. However, the quantum counterpart of initial Bell's scheme was based on theoretical possibility of preparation of *singlet states.* At that time preparation of ensembles of high quality singlet states was totally impossible. Bell understood well that his original inequality which was derived under assumption of perfect (anti-)correlations cannot be tested experimentally. And he was happy to join Clauser, Horne, Shimony, and Holt who used a new scheme and derived CHSH-inequality[11]. It seems that in future Bell had never mentioned[10] his original inequality derived in 1964, *the original Bell inequality.* The CHSH-scheme is not based on consideration of precise (anti-)correlations. It provides the possibility of performing experiment, even without clean technology for preparing singlet states.

Experimenting with the CHSH-inequality[11] and inequalities based on the same scheme[12,13,14,15] was extremely stimulating for development of quantum technologies. It was also one of biggest challenges for experimenters in history of physics. Therefore it is impossible underestimate the value of the CHSH-type inequalities for physics. However, we have to be honest and say explicitly:

The CHSH-inequality and other inequalities which are not based on precise correlations have nothing to do with the EPR framework and the Einstein-Bohr debate.

Hence, statements as "Closing the Door on Einstein and Bohr's Quantum Debate", see paper[4], are not justified. To close this door, the original Bell inequality[9] as to be tested. Nowadays the experimental technology is essentially more advance than in Bell's time. In particular, very clean ensembles of singlet states can be prepared. Photo-detectors of high efficiency were already used in quantum experiments. This makes the experiment on violation of the original Bell inequality at least less impossible than in Bell's time, see article[16] for detailed analyzing the interplay between detection efficiency and the singlet state preparation. Obviously, such an experiment is even bigger challenge for experimental physics than the previous experiments on violations of inequalities based on the CHSH-scheme.

3. Would be Bohr happy with nonlocal "closing the door" in his debate with Einstein?

From reading Bohr [17] and philosophers who put tremendous efforts to clarifying Bohr's views, e.g., book[18], we understand that for him quantum mechanics (QM) is a local theory. In particular, he did not explore the nonlocality alternative in his reply to Einstein[19]. It is practically unknown that Bohr also had his own notion of an element of reality known as *phenomenon*. And this is a local notion, see paper[20].

Hence, the talks of people claiming they are Copenhagenists and at the same time speaking about quantum nonlocality are really misleading. They should honestly reject the Copenhagen interpretation and explicitly say that not only Einstein, but also Bohr and other members of the Copenhagen school were wrong, because they were sure in quantum locality.

4. Incompatibility or nonlocality?

By Bohr *the complementarity principle* is the fundamental principle of QM. Experimentally this principle is expressed in existence of *incompatible observables.* Such observables cannot be jointly measured. Historically Heisenberg's uncertainty relation for the position and momentum observables was Bohr's starting point. In his reply to the EPR-argument[19] Bohr emphasized the role of the complementarity principle. By him the EPR-argument does not bring anything new to quantum foundations. He stressed that the complementarity principle is about the position and momentum of *a single particle.* And quantum interference experiments, as the two slit experiment, demonstrate incompatibility of these observables. Bohr did not find anything new in the EPR-argument. For him, quantum interference is the basic mystery of quantum mechanics. There is no need in "additional mysteries" such as quantum nonlocality.

The CHSH-equality and other Bell type inequalities which are not based on precise correlations are just additional statistical tests for the complementarity principle. In the CHSH-scheme, there are considered four observables A_i and $B_i, i = 1, 2$, such that $[A_i, B_j] = 0$. Hence, pairs A_i, B_j can be measured jointly and the corresponding correlations $\langle A_i B_j \rangle$ can be calculated. They violate the CHSH-inequality for some state if and only if $[A_1, A_2] \neq 0$ and $[B_1, B_2] \neq 0$, i.e., the A-observables as well as the B-observables are incompatible, cf. with Bohr's reply[19] to EPR paper[6]. This incompatibility is crucial in the CHSH framework.

For Bohr, violation of the CHSH-inequality is just a tricky form of expression of interference between projections of spin or polarization on different axes.

5. EPR framework: a loophole from subquantum world to quantum experiment

Bohr's reply to Einstein is often commented as unclear and misleading. And there is a point. We can wonder: How can complementarity help in explanation of the perfect correlations? In no way! But, for Bohr, there was no need in "explanation" of their origin. QM is an operational formalism concerning measurement outputs. The formalism predicts the existence of the EPR-states. And, for Bohr, this is the final point of the scientific treatment of this problem.

However, these correlations are intriguing and some people are seeking their explanation. Of course, such an explanation can be generated only in some subquantum theory. What is the key point of the EPR-argument? This is coupling of elements of such a hypothetical subquantum theory with measurement outputs, the elements of reality with outputs of measurements for the EPR-states. In principle, this coupling can be used as a loophole in the Copenhagen doctrine. One can try to test the predictions of hypothetical subquantum theories. But such tests are meaningful only for the EPR-states as, e.g., the singlet state.

6. Can the CHSH-scheme be used to test experimentally hidden variable theories?

I claim that the answer to this question is negative. Here "experimentally" is the crucial word. As was repeatedly pointed out, the CHSH-inequality is not about precise correlations. Therefore we cannot use the EPR loophole from the subquantum world to quantum experiment. *There is no reason to assume that subquantum correlations expressed mathematically in terms of hidden variables coincide with the experimental correlations predicted by QM.* The correlations given by integrals with respect to the distribution of hidden variables satisfy the CHSH-inequality, but generally they have no relation to correlations obtained in experiment. In fact, this was De Broglie's viewpoint, see works[21],[8]. This viewpoint match the Bild conception of scientific theory which elaborated by Hertz and Boltzmann, see preprint[22].

7. Concluding remarks

The main impact of experimental testing for CHSH-like inequalities is demonstration that correlations predicted by QM can be preserved for very long distances. The only foundational impact of such tests is the confirmation of Bohr's complementarity principle. Such inequalities and tests on their violation cannot be used for testing hypothetical subquantum theories with hidden variables. It seems that only the original Bell inequality can be used for such a purpose. Until a loophole free test for the latter will be performed, the statements as "Bohr was right and Einstein was wrong" or "Closing the door on Einstein and Bohr's quantum debate" are not justified.

References

1. B. Hensen et al., Experimental loophole-free violation of a Bell inequality using entangled electron spins separated by 1.3 km, *Nature* **526**, 682 (2015).
2. M. Giustina et al., A significant-loophole-free test of Bell's theorem with entangled photons, *Phys. Rev. Lett.* **115**, 250401 (2015).
3. L. K. Shalm et al., A strong loophole-free test of local realism, *Phys. Rev. Lett.* **115**, 250402 (2015).
4. A. Aspect, Closing the door on Einstein and Bohr's quantum debate, *Physics* **8**, 123 (2015).
5. H. Wiseman, Quantum physics: Death by experiment for local realism, *Nature* **526**, 649 (2015).
6. A. Einstein, B. Podolsky, N. Rosen, Can quantum-mechanical description of physical reality be considered complete? *Phys. Rev.* **47** (10), 777 (1935).
7. M. Kupczynski, Can Einstein with Bohr debate on quantum mechanics be closed? *Phil. Trans. Royal Soc.* A **375**, N 2106, 2016039 (2017).
8. A. Khrennikov, After Bell, *Fortschritte der Physik (Progress in Physics)* **65**, N 6-8, 1600014 (2017).
9. J. Bell, On the Einstein-Podolsky-Rosen paradox, *Physics* **1**, 195 (1964).
10. J. Bell, *Speakable and Unspeakable in Quantum Mechanics* (Cambridge Univ. Press, Cambridge, 1987).
11. J. F. Clauser, M. A. Horne, A. Shimony, and R. A . Holt Proposed experiment to test local hidden-variable theories. *Phys. Rev. Lett.* **23** (15), 880 (1969).
12. J. F. Clauser and M. A. Horne, Experimental consequences of objective local theories, *Phys. Rev.* D **10**, 526 (1974).
13. J. F. Clauser and A. Shimony, Bell's theorem. Experimental tests and implications, *Rep. Prog. Phys.* **41**, 1881 (1978).
14. Ph. H. Eberhard, Background level and counter efficiencies required for a loophole-free Einstein-Podolsky-Rosen experiment, *Phys. Rev.* A **47**, 477 (1993).
15. A. Khrennikov, S. Ramelow, R. Ursin, B. Wittmann, J. Kofler, I. Basieva, On

the equivalence of the Clauser-Horne and Eberhard inequality based tests, *Physica Scripta* **T163** 014019 (2014).

16. A. Khrennikov and I. Basieva, Towards experiments to test violation of the original Bell inequality. *Entropy* **20**(4), 280 (2018).

17. N. Bohr, The philosophical writings of Niels Bohr, 3 vols. (Ox Bow Press, Woodbridge Conn., 1987).

18. A. Plotnitsky, *Reading Bohr: Physics and philosophy* (Springer, Dordrecht, 2006).

19. N. Bohr, Can quantum-mechanical description of physical reality be considered complete? Phys. Rev. **48**, 696 (1935).

20. A. Plotnitsky and A. Khrennikov, Reality without realism: On the ontological and epistemological architecture of quantum mechanics, *Found. Phys.* **45**, N 10, 1269 (2015).

21. L. De Broglie, *The Current Interpretation of Wave Mechanics: a Critical Study* (Elsevier, 1964).

22. A. Khrennikov, Hertz's viewpoint on quantum theory. arXiv:1807.06409 [quant-ph].

Quantum Bio-Informatics VI
© 2020 World Scientific Publishing Co. Pte. Ltd.
pp. 93–100

GENE EXPRESSION LEVEL ANALYSIS OF PROTEIN CODING GENES INCLUDING NON-CODING RNA GENES IN INTRONIC REGIONS

YOSUKE KONDO and SATORU MIYAZAKI*

*Department of Medicinal and Life Science, Faculty of Pharmaceutical Sciences,
Tokyo University of Science,
2641 Yamazaki, Noda-shi, Chiba 278-8510, Japan
* E-mail: smiyazak@rs.noda.tus.ac.jp*

In 2003, the Human Genome Project was completed. And almost all gene locations on the human genome sequence were identified. However, regulation of gene expression has been far from elucidated. Here, we report gene expression analysis of host genes, which include a non-coding RNA (ncRNA) gene in the intronic region. We consider that such an inclusiveness of genes is one of the regulatory ways of gene expression because expression of an intronic ncRNA gene can be controlled by expression of a protein-coding gene. Our results showed that gene expression levels of host genes tend to be higher than those of non-host genes. And host genes orthologous between human and mouse showed more conserved expression levels than non-host orthologous genes. These results indicate that the inclusiveness of genes should be analyzed to elucidate gene expression regulation of host genes.

Keywords: Non-coding RNA; Intron; Host gene; Gene expression.

1. Introduction

In 2003, the Human Genome Project was finished, and almost all human genome sequences were deciphered.[1] Presently, in addition to the human genome data, we can obtain a variety of complete genome sequences from public databases.[2] There are a variety of genes on the genome sequences including not only protein-coding genes but also non-coding RNA (ncRNA) genes.[3] When we look gene locations on the genome sequences carefully, we can find an ncRNA gene that is located in a protein-coding gene. Such a protein-coding gene is referred to as a host gene. Meanwhile, the ncRNA gene that is located on the intronic region of the host gene is called as an intronic ncRNA gene. Because the intronic ncRNA is produced by splicing of a precursor messenger RNA (pre-mRNA), gene expression of a host gene triggers transcription of an intronic ncRNA gene at the same time (see Fig. 1).

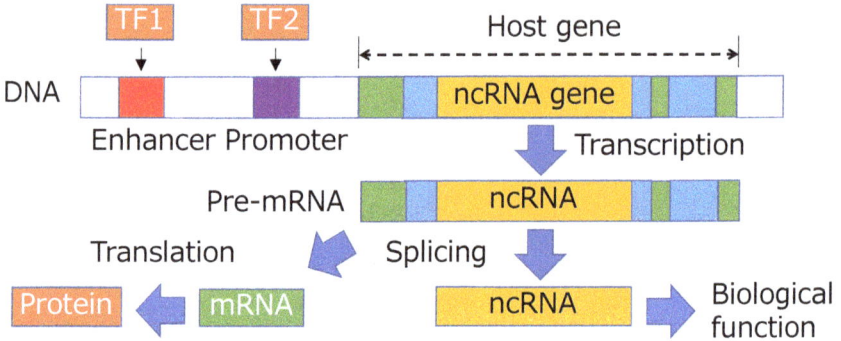

Fig. 1.   Gene expression of a host gene and its intronic ncRNA gene

Intronic ncRNAs have a variety of biological functions. For instance, if the intronic ncRNA is a micro RNA (miRNA), it is called as a mirtron,[4,5] which can regulate gene expression of a protein-coding gene by affecting the messenger RNA (mRNA). The target of a mirtron is a protein-coding gene, but some mirtron targets are their host genes.[6,7] There is a possibility that transcriptions of host genes are specially regulated by the intronic ncRNA genes. We, therefore, consider that there are some common features of gene expression of host genes. However, biological features of host genes are still unknown. In this study, we analyze gene expression levels of host genes by using an ncRNA gene database that we created previously[8] to find out some common biological features of intronic ncRNAs and their host genes.

2. Classification of ncRNA genes

To identify all host genes in genome sequences, we firstly classified ncRNA genes into three categories as follows.

(1) Intergenic
(2) Intronic
(3) Sense

This classification is based on the position of a pre-mRNA. An intergenic ncRNA gene is located on an intergenic region between protein-coding genes. An intronic ncRNA gene is located on an intronic region of a protein-coding gene. A sense ncRNA gene is located on an overlapping region between a protein-coding gene and an intergenic region. The detail is described in our previous report.[8]

3. Reconstruction of an ncRNA gene database

We downloaded gene locations in each chromosome of human and mouse genome sequences from the Ensembl genome database (Human: 1st to 22nd, X and Y of GRCh38, Mouse: 1st to 19th, X and Y of GRCm39).[2] Our database schema includes tables to store all the data downloaded from the Ensembl genome database (Release 87). Then, gene locations of host genes and intronic ncRNA genes were stored into our database. Our database also contains orthologous genes downloaded from Ensembl BioMart.[9] The orthologous genes were divided into three types such as one-to-one, one-to-many and many-to-many. The detail is described in our previous report.[8]

Our database contains 19,961 protein coding genes and 3,596 intronic ncRNA genes in human. The intronic human ncRNA genes are included in 2,691 host genes. Meanwhile, our database contains 22,050 protein coding genes and 2,197 intronic ncRNA genes in mouse. The intronic ncRNA genes are included in 1,633 host genes. These results show that some host genes include two or more ncRNA genes in the intronic regions.

4. Gene expression level analysis of host genes

We used next-generation sequencing data downloaded from Expression Atlas (https://www.ebi.ac.uk/gxa/home)[10] to compare expression levels of host genes. We found that E-MTAB-3716 and E-MTAB-3718 are useful to compare between tissues (cerebellum, heart, liver, kidney and testis) and between human and mouse in gene expression levels. These data are baseline expression levels that are obtained from healthy cells in several tissues.[11] We downloaded the expression levels based on transcript per million (TPM).[12]

We divided protein-coding genes into non-host and host genes. Fig. 2 and Fig. 3 show that distributions of expression levels of non-host and host genes in each tissue in human and mouse, respectively. The distributions in cerebellum were based on 14,283 non-host and 2,496 host genes that we obtained from the RNA-seq data. We calculated logarithms of gene expression levels (log(tpm)). In Fig. 2 and Fig. 3, the most frequent values of the numbers of genes in each tissue are between 0 to 2. The comparisons between non-host and host genes show that the numbers of host genes are

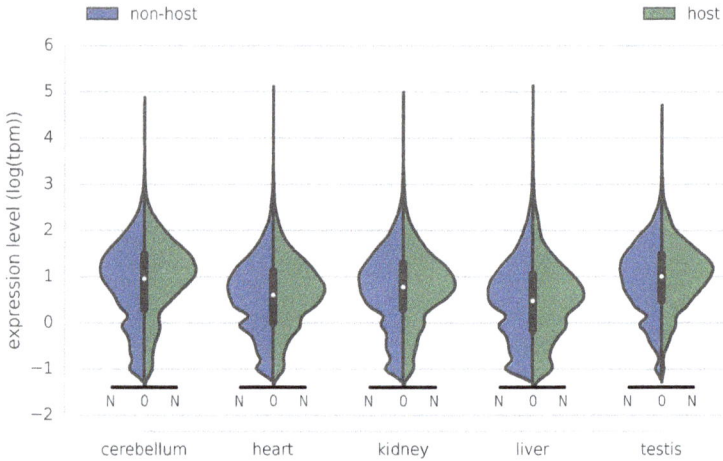

Fig. 2. Distributions of gene expression levels of protein-coding genes in human. The horizontal line in each tissue shows the number of genes.

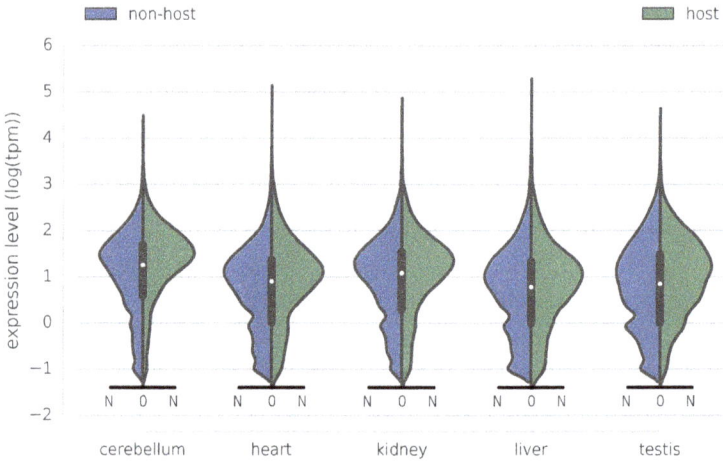

Fig. 3. Distributions of gene expression levels of protein-coding genes in mouse. The horizontal line in each tissue shows the number of genes.

larger than those of non-host genes at the most frequent values. On the other hand, there are some smaller peaks around 0 and −1 in host genes than non-host genes. These peaks show that the numbers of non-host genes are larger than those of host genes. These results indicate that expression levels of host genes may have a kind of a characteristic feature.

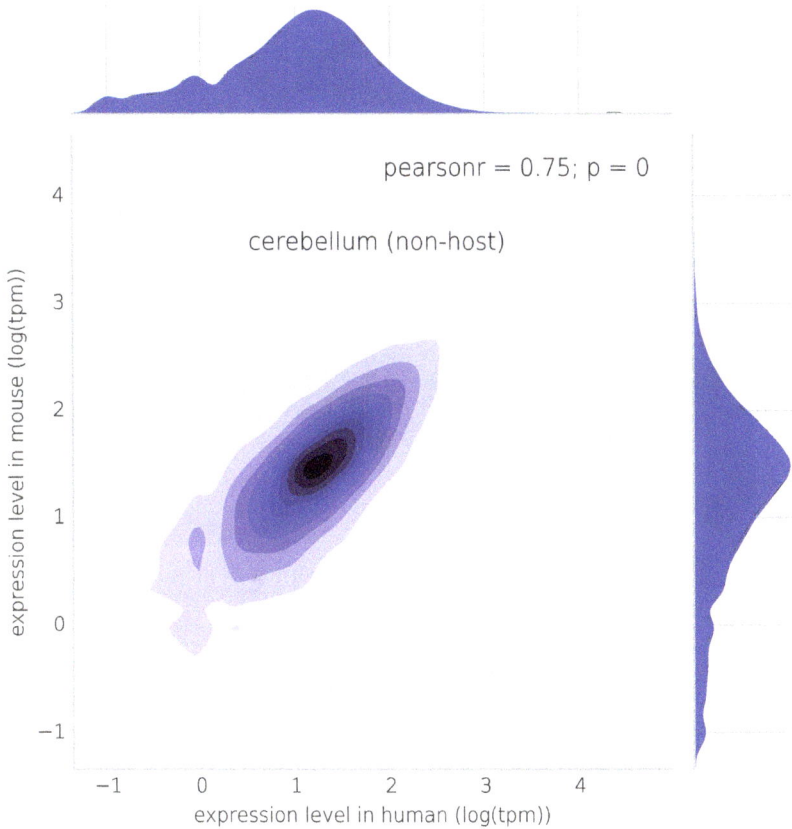

Fig. 4. Distributions of gene expression levels of non-host genes orthologous between human and mouse

There were 13,082 orthologous non-host genes and 499 orthologous host genes in cerebellum. Fig. 4 and Fig. 5 show that expression levels of non-host and host genes orthologous between human and mouse have linear relationships. The correlation coefficient of the non-host genes in cerebellum was 0.75, and that of the host genes was 0.79. The result shows that host genes have a larger value of correlation. The correlation coefficients of non-host genes in heart, kidney, liver and testis were 0.74, 0.71, 0.74 and 0.63, respectively. Meanwhile, those of host genes in each tissue were 0.81, 0.80, 0.79 and 0.74, respectively. These results show that the correlation coefficients of host genes tend to be larger than those of non-host genes

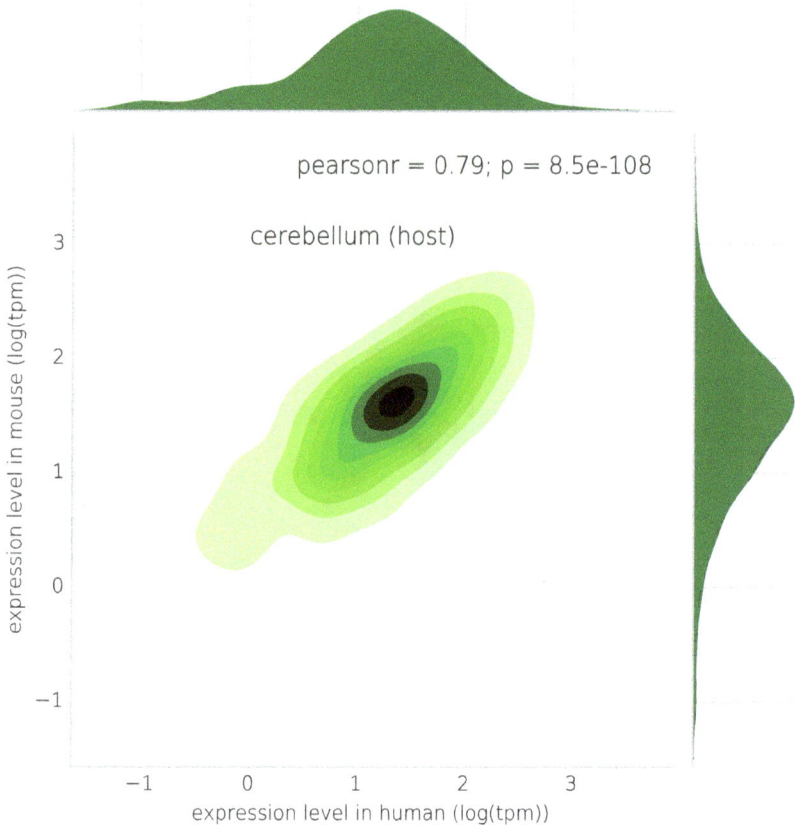

Fig. 5.   Distributions of gene expression levels of host genes orthologous between human and mouse

Therefore, expression levels of host genes orthologous between human and mouse tend to be more conserved than non-host orthologous genes.

5. Conclusions and future works

Our results showed that host genes tend to be higher expression levels than non-host genes. This biological feature may be involved in biological functions of host genes.

In future works, we will perform functional analysis by dividing genes in expression levels. And we will analyze transcriptional regulatory regions of host genes from transcription factors and their binding regions (cis-elements).

References

1. F. Collins, E. Lander, J. Rogers, R. Waterston and I. H. G. S. Conso, *Nature* **431**, 931(OCT 21 2004).
2. T. Hubbard, D. Barker, E. Birney, G. Cameron, Y. Chen, L. Clark, T. Cox, J. Cuff, V. Curwen, T. Down, R. Durbin, E. Eyras, J. Gilbert, M. Hammond, L. Huminiecki, A. Kasprzyk, H. Lehvaslaiho, P. Lijnzaad, C. Melsopp, E. Mongin, R. Pettett, M. Pocock, S. Potter, A. Rust, E. Schmidt, S. Searle, G. Slater, J. Smith, W. Spooner, A. Stabenau, J. Stalker, E. Stupka, A. Ureta-Vidal, I. Vastrik and M. Clamp, *Nucleic Acids Res* **30**, 38(JAN 1 2002).
3. S. Djebali, C. A. Davis, A. Merkel, A. Dobin, T. Lassmann, A. Mortazavi, A. Tanzer, J. Lagarde, W. Lin, F. Schlesinger, C. Xue, G. K. Marinov, J. Khatun, B. A. Williams, C. Zaleski, J. Rozowsky, M. Roeder, F. Kokocinski, R. F. Abdelhamid, T. Alioto, I. Antoshechkin, M. T. Baer, N. S. Bar, P. Batut, K. Bell, I. Bell, S. Chakrabortty, X. Chen, J. Chrast, J. Curado, T. Derrien, J. Drenkow, E. Dumais, J. Dumais, R. Duttagupta, E. Falconnet, M. Fastuca, K. Fejes-Toth, P. Ferreira, S. Foissac, M. J. Fullwood, H. Gao, D. Gonzalez, A. Gordon, H. Gunawardena, C. Howald, S. Jha, R. Johnson, P. Kapranov, B. King, C. Kingswood, O. J. Luo, E. Park, K. Persaud, J. B. Preall, P. Ribeca, B. Risk, D. Robyr, M. Sammeth, L. Schaffer, L.-H. See, A. Shahab, J. Skancke, A. M. Suzuki, H. Takahashi, H. Tilgner, D. Trout, N. Walters, H. Wang, J. Wrobel, Y. Yu, X. Ruan, Y. Hayashizaki, J. Harrow, M. Gerstein, T. Hubbard, A. Reymond, S. E. Antonarakis, G. Hannon, M. C. Giddings, Y. Ruan, B. Wold, P. Carninci, R. Guigo and T. R. Gingeras, *Nature* **489**, 101(SEP 6 2012).
4. E. Berezikov, W.-J. Chung, J. Willis, E. Cuppen and E. C. Lai, *Molecular Cell* **28**, 328(OCT 26 2007).
5. H. J. Curtis, C. R. Sibley and M. J. A. Wood, *Wiley Interdisciplinary Reviews-RNA* **3**, 617(SEP-OCT 2012).
6. J. Qian, R. Tu, L. Yuan and W. Xie, *Experimental Cell Res* **344**, 183(JUN 10 2016).
7. A. Steiman-Shimony, O. Shtrikman and H. Margalit, *RNA* **24**, 991(AUG 2018).
8. Y. Kondo, C. Hayashi and S. Miyazaki, *Journal of Melecular and Genetic Medicine* **11**, p. 2(OCT 26 2017).
9. R. J. Kinsella, A. Kaehaeri, S. Haider, J. Zamora, G. Proctor, G. Spudich, J. Almeida-King, D. Staines, P. Derwent, A. Kerhornou, P. Kersey and P. Flicek, *Database-The Journal of Biological Databases and Curation* (2011).
10. R. Petryszak, T. Burdett, B. Fiorelli, N. A. Fonseca, M. Gonzalez-Porta, E. Hastings, W. Huber, S. Jupp, M. Keays, N. Kryvych, J. McMurry, J. C. Marioni, J. Malone, K. Megy, G. Rustici, A. Y. Tang, J. Taubert,

E. Williams, O. Mannion, H. E. Parkinson and A. Brazma, *Nucleic Acids Res* **42**, D926(JAN 2014).

11. D. Brawand, M. Soumillon, A. Necsulea, P. Julien, G. Csardi, P. Harrigan, M. Weier, A. Liechti, A. Aximu-Petri, M. Kircher, F. W. Albert, U. Zeller, P. Khaitovich, F. Gruetzner, S. Bergmann, R. Nielsen, S. Paeaebo and H. Kaessmann, *Nature* **478**, 343+(OCT 20 2011).

12. G. P. Wagner, K. Kin and V. J. Lynch, *Theory in Biosciences* **131**, 281(DEC 2012).

Quantum Bio-Informatics VI
© 2020 World Scientific Publishing Co. Pte. Ltd.
pp. 101–112

ON QUANTUM QUADRATIC STOCHASTIC OPERATORS WITH KADISON-SCHWARZ PROPERTY

FARRUKH MUKHAMEDOV

Department of Computational & Theoretical Sciences
Faculty of Science, International Islamic University Malaysia
P.O. Box, 141, 25710, Kuantan
Pahang, Malaysia
E-mail: far75m@yandex.ru; farrukh_m@iium.edu.my

In the present paper we review the recent development on Kadison-Schwarz type quantum quadratic stochastic operators acting on the algebra of 2×2 matrices.

Keywords: Quantum quadratic stochastic operator; positive; Kadison-Schwarz.

1. Introduction

It is known that entanglement is one of the essential features of quantum physics and is fundamental in modern quantum technologies [34]. One of the central problems in the entanglement theory is the discrimination between separable and entangled states. There are many papers devoted to find a given state is separable (see [16]). The most general approach to characterize quantum entanglement uses a notion of an entanglement witness [17,8]. This uses the positivity of some mappings [2,7,9,15]. Therefore, it would be interesting to find some conditions for the positivity of given mappings [22,23,24]. In the literature the most tractable maps, the completely positive mapping, have proved to be of great importance in the structure theory of C*-algebras (see [6,22,35,36,41]). It is therefore of interest to study conditions stronger than positivity, but weaker than complete positivity. Such a condition is called a *Kadison-Schwarz (KS) property*. Note that KS-operators need not be completely positive. In [38] relations between n-positivity of a map ϕ and the KS property of certain map is established (see also [4]). Some ergodic properties of the Kadison-Schwarz maps were investigated in [14,39]. In [29] we have described bistochastic KS-operators from $M_2(\mathbb{C})$ to itself. Recently, in [10], we have introduced a concept of Kadison-Schwarz divisible dynamical maps. It turns out that it is a natural generalization of the well known CP-divisibility which characterizes quantum Markovian evolution. It is proved that Kadison-Schwarz divisible maps are fully characterized in terms of time-local dissipative generators.

In the present paper we review the recent development on Kadison-Schwarz type quantum quadratic stochastic operators acting on the algebra of 2×2 matrices $\mathbb{M}_2(\mathbb{C})$.

2. Preliminaries

Let $M_n(\mathbb{C})$ be the algebra of $n \times n$ matrices over the complex field $\mathbb{C}$. Let A and B be two C^*-algebras with unit. Recall that a linear mapping $\Phi : A \to B$ is called

   (i) *morphism* if $\Phi(x^*) = \Phi(x)^*$ for all $x \in A$;

   (i) *positive* if $\Phi(x) \geq 0$ whenever $x \geq 0$;

   (ii) *unital* if $\Phi(\mathbf{1}) = \mathbf{1}$;

   (iii) *n-positive* if the mapping $\Phi_n : M_n(A) \to M_n(B)$ defined by $\Phi_n(a_{ij}) = (\Phi(a_{ij}))$ is positive. Here $M_n(A)$ denotes the algebra of $n \times n$ matrices with A-valued entries;

   (iv) *completely positive* if it is n-positive for all $n \in \mathbb{N}$;

   (vii) *Kadison-Schwarz operator (KS-operator)*, if one has

$$\Phi(x)^* \Phi(x) \leq \Phi(x^* x) \quad \text{for all} \ \ x \in A. \tag{1}$$

It is well known [37] that the completely positivity can be formulated as follows: for any two collections $a_1, \cdots, a_n \in A$ and $b_1, \cdots, b_n \in B$ the following relation holds

$$\sum_{i,j=1}^{n} b_i^* T(a_i^* a_j) b_j \geq 0. \tag{2}$$

It is well–known (cf. [37]) that $\sup_n \|T_n\| = T(\mathbf{1})$ for completely positive maps. It is clear that completely positivity of T implies its positivity. In general, converse, it is not true.

Note that every unital completely positive map is KS-operator, and a famous result of Kadison states that any positive unital map satisfies (1) for every self-adjoint elements.

There are several connections between CP and KS-operators. Namely, let $\Phi : A \to B$ be a given unital mapping. For a positive invertible $a \in A$ let us define

$$\Phi_a(x) = \Phi(a^2)^{-1/2} \Phi(axa) \Phi(a^2)^{-1/2}.$$

Theorem 2.1. [38] *Let $\Phi : A \to B$ be a positive unital map. Then Φ is n-positive if and only if $(\Phi_a)_n$ is KS-operator for all positive invertible $a \in A$.*

By $\mathcal{KS}(A,B)$ we denote the set of all KS-operators mapping from A to B.

Theorem 2.2. [30] *Let A, B and C be C^*-algebras. The following statements hold:*

(i) let $\Phi, \Psi \in \mathcal{KS}(A,B)$, then for any $\lambda \in [0,1]$ the mapping $\Gamma_\lambda = \lambda\Phi+(1-\lambda)\Psi$ belongs to $\mathcal{KS}(A,B)$. This means $\mathcal{KS}(A,B)$ is convex;
(ii) let U,V be unitaries in A and B, respectively, then for any $\Phi \in \mathcal{KS}(A,B)$ the mapping $\Psi_{U,V}(x) = U\Phi(VxV^*)U^*$ belongs to $\mathcal{KS}(A,B)$;
(iii) let $\Phi \in \mathcal{KS}(A,B)$, $\Psi \in \mathcal{KS}(B,C)$, then $\Psi \circ \Phi \in \mathcal{KS}(A,C)$.

Let $B(H)$ be the set of linear bounded operators from a complex Hilbert space H to itself. By $B(H) \otimes B(H)$ we mean tensor product of $B(H)$ into itself. By $S(B(H))$ we denote the set of all states defined on $B(H)$. Let $\Delta : B(H) \to B(H) \otimes B(H)$ be a linear operator. Let $U : B(H) \otimes B(H) \to B(H) \otimes B(H)$ be a linear operator such that $U(x \otimes y) = y \otimes x$ for all $x,y \in \mathbb{M}_2(\mathbb{C})$.

Definition 2.1. A linear operator $\Delta : B(H) \to B(H) \otimes B(H)$ is said to be a *quantum quadratic stochastic operator (q.q.s.o)* if it is unital, positive and *symmetric*, i.e. $U\Delta = \Delta$.

A state $h \in S(B(H))$ is called *a Haar state* for a q.q.s.o. Δ if for every $x \in B(H)$ one has

$$(h \otimes id) \circ \Delta(x) = (id \otimes h) \circ \Delta(x) = h(x)\mathbf{1}. \tag{3}$$

Remark 2.1. In [11] we first introduced a notion of q.q.s.o.. Certain ergodic properties of such kind of operators have been investigated in [12,25]. For recent reviews on this topic we refer to [13].

One can see that each q.q.s.o. Δ defines a conjugate operator $\Delta^* : (B(H) \otimes B(H))^* \to B(H)^*$ by

$$\Delta^*(f)(x) = f(\Delta x), \; f \in (B(H) \otimes B(H))^*, \; x \in B(H).$$

One can define an operator V_Δ by

$$V_\Delta(\varphi) = \Delta^*(\varphi \otimes \varphi), \; \varphi \in B(H)^*.$$

Recall that a linear operator $\Delta : B(H) \to B(H) \otimes B(H)$ is said to be a *quasi quantum quadratic operator (quasi q.q.o)* if it is unital, $*$-preserving (i.e. $\Delta(x^*) = \Delta(x)^*$, for all $x \in B(H)$) and

$$V_\Delta(\varphi) \in B(H)_+^* \quad \text{whenever } \varphi \in B(H)_+^*; \tag{4}$$

Note that from unitality of Δ we conclude that for any quasi q.q.o. V_Δ maps $S(B(H))$ into itself. In some literature operator V_Δ is called *quadratic convolution* (see for example [18]). In [27,28,31] we have investigated quasi q.q.o. on $\mathbb{M}_2(\mathbb{C})$. Furthermore, in [29,33] certain dynamical properties of V_Δ associated with q.q.o. defined on $\mathbb{M}_2(\mathbb{C})$ are investigated.

3. Quantum quadratic stochastic operators on $\mathbb{M}_2(\mathbb{C})$

By $\mathbb{M}_2(\mathbb{C})$ be an algebra of 2×2 matrices over complex field $\mathbb{C}$. In this section we are going to describe quantum quadratic operators on $\mathbb{M}_2(\mathbb{C})$ as well as find necessary conditions for such operators to satisfy the Kadison-Schwarz property.

Recall [5] that the identity and Pauli matrices $\{\mathbf{1}, \sigma_1, \sigma_2, \sigma_3\}$ form a basis for $\mathbb{M}_2(\mathbb{C})$, where

$$\sigma_1 = \begin{pmatrix} 0 & 1 \\ 1 & 0 \end{pmatrix} \quad \sigma_2 = \begin{pmatrix} 0 & -i \\ i & 0 \end{pmatrix} \quad \sigma_3 = \begin{pmatrix} 1 & 0 \\ 0 & -1 \end{pmatrix}.$$

In this basis every matrix $x \in \mathbb{M}_2(\mathbb{C})$ can be written as $x = w_0 \mathbf{1} + \mathbf{w}\sigma$ with $w_0 \in \mathbb{C}$, $\mathbf{w} = (w_1, w_2, w_3) \in \mathbb{C}^3$, here $\mathbf{w}\sigma = w_1\sigma_1 + w_2\sigma_2 + w_3\sigma_3$. In what follows, we frequently use notation $\overline{\mathbf{w}} = (\overline{w_1}, \overline{w_2}, \overline{w_3})$.

Lemma 3.1. [40] *The following assertions hold true:*

(a) x *is self-adjoint iff* $w_0, \mathbf{w}$ *are reals;*

(b) $\text{Tr}(x) = 1$ *iff* $w_0 = 0.5$, *here* Tr *is the trace of a matrix* x;

(c) $x > 0$ *iff* $\|\mathbf{w}\| \leq w_0$, *where* $\|\mathbf{w}\| = \sqrt{|w_1|^2 + |w_2|^2 + |w_3|^2}$;

(d) *A linear functional* φ *on* $\mathbb{M}_2(\mathbb{C})$ *is a state iff it can be represented by*

$$\varphi(w_0\mathbf{1} + \mathbf{w}\sigma) = w_0 + \langle \mathbf{w}, \mathbf{f} \rangle, \tag{5}$$

where $\mathbf{f} = (f_1, f_2, f_3) \in \mathbb{R}^3$ *such that* $\|\mathbf{f}\| \leq 1$. *Here as before* $\langle \cdot, \cdot \rangle$ *stands for the scalar product in* $\mathbb{C}^3$.

(e) *A state* φ *is a pure if and only if* $\|\mathbf{f}\| = 1$. *So pure states can be seen as the elements of unit sphere in* $\mathbb{R}^3$.

In the sequel we shall identify a state with a vector $\mathbf{f} \in \mathbb{R}^3$. By τ we denote a normalized trace, i.e. $\tau(x) = \frac{1}{2}\operatorname{Tr}(x)$, $x \in M_2(\mathbb{C})$.

It is clear that the system $\{\mathbf{1} \otimes \mathbf{1}, \mathbf{1} \otimes \sigma_i, \sigma_j \otimes \mathbf{1}, \sigma_i \otimes \sigma_j\}_{i,j=1}^3$ forms a basis in $M_2(\mathbb{C}) \otimes M_2(\mathbb{C})$.

Let $\Delta : M_2(\mathbb{C}) \to M_2(\mathbb{C}) \otimes M_2(\mathbb{C})$ be a q.q.s.o. Then we write the operator Δ in terms of a basis of $M_2(\mathbb{C}) \otimes M_2(\mathbb{C})$, formed by the Pauli matrices, as follows

$$\Delta\mathbf{1} = \mathbf{1} \otimes \mathbf{1};$$

$$\Delta(\sigma_i) = b_i(\mathbf{1} \otimes \mathbf{1}) + \sum_{j=1}^3 b_{ji}^{(1)}(\mathbf{1} \otimes \sigma_j) + \sum_{j=1}^3 b_{ji}^{(2)}(\sigma_j \otimes \mathbf{1})$$

$$+ \sum_{m,l=1}^3 b_{ml,i}(\sigma_m \otimes \sigma_l), \tag{6}$$

where $i = 1, 2, 3$.

Hence using Def. 2.1 we obtain $b_{uv,k} = b_{vu,k}$ and $b_{u,k} := b_{u,k}^{(1)} = b_{u,k}^{(2)}$ therefore we have

$$\Delta(\sigma_k) = b_k \mathbf{1} + \sum_{u,v=1}^3 b_{uv,k} \sigma_u \otimes \sigma_u + \sum_{u=1}^3 b_{u,k}(\sigma_u \otimes \mathbf{1} + \mathbf{1} \otimes \sigma_u). \tag{7}$$

In general, a description of positive operators is one of the main problems of quantum information. In the literature most tractable maps are positive and trace-preserving ones, since such maps arise naturally in quantum information theory (see [19,20,34,40]). Therefore, in the sequel we shall restrict ourselves to the trace preserving q.q.s.o., i.e. $\tau \otimes \tau \circ \Delta = \tau$. So, we would like to describe all such kind of maps.

Proposition 3.1. *Let $\Delta : M_2(\mathbb{C}) \to M_2(\mathbb{C}) \otimes M_2(\mathbb{C})$ be a trace preserving q.q.s.o., then in (6) one has $b_j = 0$, and b_{ij} and $b_{ij,k} = b_{ji,k}$ are reals for every $i, j, k \in \{1, 2, 3\}$. Moreover, Δ has the following form:*

$$\Delta(x) = w_0 \mathbf{1} \otimes \mathbf{1} + \mathbf{Bw} \cdot \sigma \otimes \mathbf{1} + \mathbf{1} \otimes \mathbf{Bw} \cdot \sigma + \sum_{m,l=1}^3 \langle \mathbf{b}_{ml}, \overline{\mathbf{w}} \rangle \sigma_m \otimes \sigma_l, \tag{8}$$

where $x = w_0 + \mathbf{w}\sigma$, $\mathbf{b}_{ml} = (b_{ml,1}, b_{ml,2}, b_{ml,3})$, and $\mathbf{B} = (b_{ij})_{i,j=1}^3$. Here as before $\langle \cdot, \cdot \rangle$ stands for the standard scalar product in $\mathbb{C}^3$.

One can rewrite (8) as follows

$$\Delta(x) = \lambda \Delta_1(x) + (1 - \lambda)\Delta_2(x), \tag{9}$$

where

$$\Delta_1(x) = w_0 \mathbf{1} \otimes \mathbf{1} + \frac{1}{\lambda} \sum_{m,l=1}^{3} \langle \mathbf{b}_{ml}, \overline{\mathbf{w}} \rangle \sigma_m \otimes \sigma_l, \tag{10}$$

$$\Delta_2(x) = w_0 \mathbf{1} \otimes \mathbf{1} + \frac{1}{1-\lambda} \left(\mathbf{Bw} \cdot \sigma \otimes \mathbf{1} + \mathbf{1} \otimes \mathbf{Bw} \cdot \sigma \right). \tag{11}$$

A q.q.s.o. of the form (10) (resp. (11)) is called *simple* (resp. *non-simple*). So, any q.q.s.o. is a convex combination of simple and non-simple q.q.s.o. In the sequel we are going to investigate simple and non-simple q.q.s.o. one by one.

Remark 3.1. Note that if τ is a Haar state for Δ (see (3)), then Δ can be written as follows

$$\Delta(x) = w_0 \mathbf{1} \otimes \mathbf{1} + \sum_{m,l=1}^{3} \langle \mathbf{b}_{ml}, \overline{\mathbf{w}} \rangle \sigma_m \otimes \sigma_l, \tag{12}$$

which means that Δ is non-simple.

Let us turn to the positivity of Δ. Given a vector $\mathbf{f} = (f_1, f_2, f_3) \in \mathbb{R}^3$ put

$$\beta(\mathbf{f})_{ij} = \sum_{k=1}^{3} b_{ki,j} f_k. \tag{13}$$

Define a matrix $\mathbf{b}(\mathbf{f}) = (\beta(\mathbf{f})_{ij})_{ij=1}^{3}$, and by $\|\mathbf{b}(\mathbf{f})\|$ we denote its norm associated with Euclidean norm in $\mathbb{R}^3$.

Given a state φ by E_φ we denote the canonical conditional expectation defined by $E_\varphi(x \otimes y) = \varphi(x)y$, where $x, y \in \mathbb{M}_2(\mathbb{C})$.

In the sequel by $\mathbf{S}$ we denote the unit ball in $\mathbb{R}^3$, i.e.

$$\mathbf{S} = \{ \mathbf{p} = (p_1, p_2, p_3) \in \mathbb{R}^3 : p_1^2 + p_2^2 + p_3^2 \leq 1 \}.$$

Let us denote

$$\|\|\mathbf{b}\|\| = \sup_{\mathbf{f} \in S} \|\mathbf{b}(\mathbf{f})\|.$$

Proposition 3.2. *Let Δ be a trace preserving q.q.s.o., then one has*

$$|\langle \mathbf{Bw}, \mathbf{f} \rangle| \leq 1, \quad \|(\mathbf{B} + \mathbb{B}(\mathbf{f}))\mathbf{w}\| \leq 1 + |\langle \mathbf{Bw}, \mathbf{f} \rangle|, \tag{14}$$

for all $\mathbf{w}, \mathbf{f} \in \mathbf{S}$.

Corollary 3.1. *Let Δ be a q.q.s.o. with Haar state τ, then one has $\|\|\mathbb{B}\|\| \leq 1$.*

Remark 3.2. Note that similar characterizations of positive maps defined on $\mathbb{M}_2(\mathbb{C})$ were considered in [24]. Characterization of completely positive mappings from $\mathbb{M}_2(\mathbb{C})$ into itself with invariant state τ was established in [40].

4. Simple Kadison-Schwarz type q.q.s.o.

In this section we are going to study simple q.q.s.o. which satisfy Kadison-Schwarz condition (1) and complete positivity.

Let $\Delta : M_2(\mathbb{C}) \to M_2(\mathbb{C}) \otimes M_2(\mathbb{C})$ be a linear operator which is given by

$$\Delta(w_0\mathbf{1} + \mathbf{w} \cdot \sigma) = w_0\mathbf{1} \otimes \mathbf{1} + \mathbf{Bw} \cdot \sigma \otimes \mathbf{1} + \mathbf{1} \otimes \mathbf{Bw} \cdot \sigma \qquad (15)$$

where $\mathbf{B}$ are linear operators on $\mathbb{C}^3$.

Clearly, Δ is unital. Now let us first find conditions when Δ is positive, i.e. q.q.s.o. This is given by the following

Theorem 4.1. *The mapping Δ given by (15) is a q.q.s.o. if and only if $\|\mathbf{B}\| \leq 1/2$.*

Now let us turn to the Kadison-Schwarz property.
Define the following mapping

$$\Phi(x) = w_0\mathbf{1} + 2\mathbf{Bw} \cdot \sigma. \qquad (16)$$

Then from (15),(16) one finds

$$\Delta(x) = \frac{1}{2}\left(\Phi(x) \otimes \mathbf{1} + \mathbf{1} \otimes \Phi(x)\right). \qquad (17)$$

From Theorem 2.2 we immediately find

Corollary 4.1. *If the mapping Φ given by (16) is a KS-operator, then Δ given by (17) is also KS-operator.*

We are interested in finding more general condition than the formulated one.

Theorem 4.2. [32] *Let Δ be a simple q.q.s.o. given by (17). If one has*

$$\|\mathbf{B}\| \leq \frac{1}{2} \qquad (18)$$

$$2\|\mathbf{B}[\mathbf{w}, \overline{\mathbf{w}}] - 2[\mathbf{Bw}, \mathbf{B}\overline{\mathbf{w}}]\|\| \leq \|\mathbf{w}\|^2 - 4\|\mathbf{Bw}\|^2. \qquad (19)$$

Then Δ is a Kadison-Schwarz operator.

We should stress that the conditions (18),(19) are sufficient to be KS-operator.

Remark 4.1. We have to stress that if Δ is KS operator, then the mapping Φ, in general, does not need to be KS-operator.

Let D be a simple q.q.s.o. given by (15). Then following [19] let us decompose the matrix $\mathbf{B}$ as follows $\mathbf{B} = \mathbf{R}\mathbf{A}$, here $\mathbf{R}$ is a rotation and $\mathbf{A}$ is a self-adjoint matrix (see [19]). Define a mapping $\Delta_{\mathbf{A}}$ as follows

$$\Delta_{\mathbf{A}}(w_0\mathbf{1} + \mathbf{w} \cdot \sigma) = w_0\mathbf{1} \otimes \mathbf{1} + \mathbf{Aw} \cdot \sigma \otimes \mathbf{1} + \mathbf{1} \otimes \mathbf{Aw} \cdot \sigma.$$

Every rotation is implemented by a unitary matrix in $M_2(\mathbb{C})$, therefore there is a unitary $U \in M_2(\mathbb{C})$ such that

$$\Delta(x) = U\Delta_{\mathbf{A}}(x)U^*, \quad x \in M_2(\mathbb{C}). \tag{20}$$

On the other hand, every self-adjoint operator $\mathbf{A}$ can be diagonalized by some unitary operator, i.e. there is a unitary $V \in M_2(\mathbb{C})$ such that $\mathbf{A} = VD_{\lambda_1,\lambda_2,\lambda_3}V^*$, where

$$D_{\lambda_1,\lambda_2,\lambda_3} = \begin{pmatrix} \lambda_1 & 0 & 0 \\ 0 & \lambda_2 & 0 \\ 0 & 0 & \lambda_3 \end{pmatrix}, \tag{21}$$

where $\lambda_1, \lambda_2, \lambda_3 \in \mathbb{R}$.

Consequently, the mapping Δ can be represented by

$$\Delta(x) = \tilde{U}\Delta_{D_{\lambda_1,\lambda_2,\lambda_3}}(x)\tilde{U}^*, \quad x \in M_2(\mathbb{C}) \tag{22}$$

for some unitary $\tilde{U}$. Due to Theorem 2.2, the mapping $\Delta_{D_{\lambda_1,\lambda_2,\lambda_3}}$ is also KS-operator. Hence, all simple q.q.s.o. with KS-property can be characterized by $\Delta_{D_{\lambda_1,\lambda_2,\lambda_3}}$ and unitaries. In what follows, for the sake of shortness, by $\Delta_{(\lambda_1,\lambda_2,\lambda_3)}$ we denote the mapping $\Delta_{D_{\lambda_1,\lambda_2,\lambda_3}}$. From Theorem 4.1 one finds that $|\lambda_k| \leq 1/2$, for all $k = 1, 2, 3$.

Next, we want to characterize KS-operators of the form $\Delta_{(\lambda_1,\lambda_2,\lambda_3)}$.

Theorem 4.3. *If*

$$(1 + 4\lambda_1^2)(3 + 4\lambda_2^2 + 4\lambda_3^2 - 4\lambda_1^2) \leq 4(1 + 8\lambda_1\lambda_2\lambda_3),$$
$$(1 + 4\lambda_2^2)(3 + 4\lambda_1^2 + 4\lambda_3^2 - 4\lambda_2^2) \leq 4(1 + 8\lambda_1\lambda_2\lambda_3),$$
$$(1 + 4\lambda_3^2)(3 + 4\lambda_1^2 + 4\lambda_2^2 - 4\lambda_3^2) \leq 4(1 + 8\lambda_1\lambda_2\lambda_3)$$

are satisfied, then $\Delta_{(\lambda_1,\lambda_2,\lambda_3)}$ *is a KS-operator.*

It is interesting to study when the operator $\Delta_{(\lambda_1,\lambda_2,\lambda_3)}$ is complete positive. The next result characterizes the complete positivity of $\Delta_{(\lambda_1,\lambda_2,\lambda_3)}$.

Theorem 4.4. *A map* $\Delta_{(\lambda_1,\lambda_2,\lambda_3)}$ *is completely positive if and only if one of the following inequalities are satisfied*

(1) $|\lambda_3| < \frac{1}{2}$,
$$4\lambda_1^2 + 4\lambda_2^2 + 4\lambda_3^2 \le 1 + 16\lambda_1\lambda_2\lambda_3,$$
$$\lambda_1^2 + \lambda_2^2 + \sqrt{\left(\lambda_1^2 + \lambda_2^2\right)^2 - 4\lambda_1\lambda_2\lambda_3 + \lambda_3^2} \le \frac{1}{2};$$
(2) $\lambda_3 = \frac{1}{2}$, $|\lambda_1| \le \frac{1}{2}$, $|\lambda_2| \le \frac{1}{2}$;
(3) $\lambda_3 = -\frac{1}{2}$, $\lambda_1 = \pm\frac{1}{2}$, $\lambda_2 = \mp\frac{1}{2}$.

5. Non-simple Kadison-Schwarz type q.q.s.o.

In this section, we are going to find some conditions for non-simple q.q.s.o. to be Kadison-Schwarz operators, meaning we consider a q.q.s.o. Δ with a Haar state τ, i.e. Δ has the following form

$$\Delta(x) = w_0 \mathbf{1} \otimes \mathbf{1} + \sum_{m,l=1}^{3} \langle \mathbf{b}_{ml}, \overline{\mathbf{w}} \rangle \sigma_m \otimes \sigma_l. \tag{23}$$

Let us introduce some notations. Given $x = w_0 + \mathbf{w}\sigma$ and a vector $\mathbf{f} \in \mathbf{S}$ we denote

$$x_{ml} = \langle \mathbf{b}_{ml}, \mathbf{w} \rangle, \quad \mathbf{x}_m = \left(\langle \mathbf{b}_{m1}, \mathbf{w} \rangle, \langle \mathbf{b}_{m2}, \mathbf{w} \rangle, \langle \mathbf{b}_{m3}, \mathbf{w} \rangle \right), \tag{24}$$

$$\alpha_{ml} = \langle \mathbf{x}_m, \mathbf{x}_l \rangle - \langle \mathbf{x}_l, \mathbf{x}_m \rangle, \quad \gamma_{ml} = [\mathbf{x}_m, \overline{\mathbf{x}_l}] + [\overline{\mathbf{x}_m}, \mathbf{x}_l], \tag{25}$$

$$\mathbf{q}(\mathbf{f}, \mathbf{w}) = \left(\langle \beta(\mathbf{f})_1, [\mathbf{w}, \overline{\mathbf{w}}] \rangle, \langle \beta(\mathbf{f})_2, [\mathbf{w}, \overline{\mathbf{w}}] \rangle, \langle \beta(\mathbf{f})_3, [\mathbf{w}, \overline{\mathbf{w}}] \rangle \right), \tag{26}$$

where $\beta(\mathbf{f})_m = (\beta(\mathbf{f})_{m1}, \beta(\mathbf{f})_{m2}, \beta(\mathbf{f})_{m3})$ and as before $\mathbf{b}_{ml} = (b_{ml,1}, b_{ml,2}, b_{ml,3})$

By π we denote mapping $\{1, 2, 3, 4\}$ to $\{1, 2, 3\}$ defined by $\pi(1) = 2, \pi(2) = 3, \pi(3) = 1, \pi(4) = \pi(1)$.

Theorem 5.1. [33] *Let* $\Delta : \mathbb{M}_2(\mathbb{C}) \to \mathbb{M}_2(\mathbb{C}) \otimes \mathbb{M}_2(\mathbb{C})$ *be a q.q.s.o. with a Haar state* τ *(see (23)). Assume that* Δ *is Kadison-Schwarz operator. Then the coefficients* $\{b_{ml,k}\}$ *satisfy the following conditions*

$$\|\mathbf{w}\|^2 \ge i \sum_{m=1}^{3} f_m \alpha_{\pi(m),\pi(m+1)} + \sum_{m=1}^{3} \|\mathbf{x}_m\|^2 \tag{27}$$

$$\left\| q(\mathbf{f}, \mathbf{w}) - i \sum_{m=1}^{3} f_m \gamma_{\pi(m),\pi(m+1)} - [\mathbf{x}_m, \overline{\mathbf{x}}_m] \right\| \leq \|\mathbf{w}\|^2 - i \sum_{k=1}^{3} f_k \alpha_{\pi(k),\pi(k+1)}$$

$$- \sum_{m=1}^{3} \|\mathbf{x}_m\|^2 \qquad (28)$$

for all $\mathbf{f} \in \mathbf{S}, \mathbf{w} \in \mathbb{C}^3$.

Remark 5.1. The provided characterization with [24] allows us to construct examples of positive or Kadison-Schwarz operators which are not completely positive.

6. An Example of non-simple q.q.s.o. which is not Kadison-Schwarz one

Let us consider the following linear operator:

$$\Delta_\varepsilon(x) = w_0 \mathbf{1} \otimes \mathbf{1} + \varepsilon w_1 \sigma_1 \otimes \sigma_1 + \varepsilon w_3 \sigma_1 \otimes \sigma_2 + \varepsilon w_2 \sigma_1 \otimes \sigma_3$$
$$+ \varepsilon w_3 \sigma_2 \otimes \sigma_1 + \varepsilon w_2 \sigma_2 \otimes \sigma_2 + \varepsilon w_1 \sigma_2 \otimes \sigma_3$$
$$+ \varepsilon w_2 \sigma_3 \otimes \sigma_1 + \varepsilon w_1 \sigma_3 \otimes \sigma_2 + \varepsilon w_3 \sigma_3 \otimes \sigma_3, \qquad (29)$$

where as before $x = w_0 \mathbf{1} + \mathbf{w}\sigma$.

Theorem 6.1. *Let* $\Delta_\varepsilon : \mathbb{M}_2(\mathbb{C}) \to \mathbb{M}_2(\mathbb{C}) \otimes \mathbb{M}_2(\mathbb{C})$ *be given by* (29). *Then the following statements hold:*

(i) Δ_ε *is a q.q.s.o. if and only if* $|\varepsilon| \leq \frac{1}{3}$;
(ii) Let $\varepsilon = \frac{1}{3}$ *then the corresponding q.q.s.o.* Δ_ε *is not a KS-operator;*
(iii) Δ_ε *is completely positive if and only if* $|\varepsilon| \leq \frac{1}{3\sqrt{3}}$.

Note that the statements (i) and (ii) were proved in [29] by means of Theorem 5.1.

Acknowledgments

The authors would like to thank Professors Noboru Watanabe for his kind invitation to "QBIC Workshop 2014" (23-25 October 2014). Moreover, the author is grateful Tokyo University of Science for kind hospitality and support.

References

1. L.Accardi, M. Ohya, Compound channels, transition expectations, and liftings, *Appl. Math. Optimization* **39**(1999) 33–59.
2. I. Bengtsson, K. Zyczkowski, *Geometry of quantum states*, Cambridge Univ. Press, 2006.
3. S. N. Bernstein, The solution of a mathematical problem related to the theory of heredity, *Uchen. Zapiski NI Kaf. Ukr. Otd. Mat.* 1924, no. 1, 83–115. (Russian)
4. S.J. Bhatt, Stinespring representability and Kadison's Schwarz inequality in non-unital Banach star algebras and applications, *Proc. Indian Acad. Sci. (Math. Sci.)*, **108** (1998), 283–303.
5. O. Bratteli, D. W. Robertson, *Operator algebras and quantum statistical mechanics*. I, Springer, New York - Heidelberg - Berlin 1979.
6. M-D. Choi, Completely positive linear maps on complex matrices, *Lin. Alg. Appl.* **10**(1975), 285–290.
7. D. Chruscinski, Quantum-correlation breaking channels, quantum conditional probability and Perron Frobenius theory, *Phys. Lett A* **377**(2013), 606–611.
8. D. Chruscinski, A class of symmetric Bell diagonal entanglement witnesses a geometric perspective, *J. Phys. A: Math. Theor.* **47**(2014), 424033.
9. D. Chruscinski, G. Sarbicki, Exposed positive maps in $M_4(C)$, *Open Systems & Infor. Dynam.* **19** (2012), 1250017.
10. D. Chruscinski, F. Mukhamedov, Dissipative generators, divisible dynamical maps and Kadison-Schwarz inequality, *Phys. Rev. A* **100**(2019), 052120.
11. N. N. Ganikhodzhaev, F. M. Mukhamedov, On quantum quadratic stochastic processes, and some ergodic theorems for such processes, *Uzb. Matem. Zh.* 1997, no. 3, 8–20. (Russian)
12. N.N. Ganikhodzhaev, F. M. Mukhamedov, Ergodic properties of quantum quadratic stochastic processes, *Izv. Math.* **65** (2000), 873–890.
13. R. Ganikhodzhaev, F. Mukhamedov, U. Rozikov, Quadratic stochastic operators and processes: results and open problems, *Inf. Dim. Anal. Quantum Probab. and Related Topics* **14**(2011) 279–335.
14. U. Groh, Uniform ergodic theorems for identity preserving Schwarz maps on W^*-algebras, *J. Operator Theory* **11** (1984), 395–404.
15. K.-C. Ha, Entangled states with strong positive partial transpose, *Phys. Rev. A* **81**(2010), 064101.
16. R. Horodecki, P. Horodecki, M. Horodecki and K. Horodecki, Quantum entanglement, *Rev. Mod. Phys.* **81**(2009), 865.
17. M. Horodecki, P. Horodecki, R. Horodecki, Separability of mixed states: Necessary and sufficient conditions, *Phys. Lett. A* **223** (1996), 1–8.
18. U. Franz, A. Skalski, On ergodic properties of convolution operators associated with compact quantum groups, *Colloq. Math.* **113** (2008), 13–23.
19. C. King, M.B. Ruskai, Minimal entropy of states emerging from noisy quantum channels. *IEEE Trans. Info. Theory* **47**, (2001) 192–209.
20. A. Kossakowski, A Class of Linear Positive Maps in Matrix Algebras, *Open Sys. & Information Dyn.* **10**(2003) 213–220.

21. Yu. I. Lyubich, *Mathematical structures in population genetics*, Springer, Berlin 1992.

22. W.A. Majewski, M. Marciniak, On a characterization of positive maps, *J. Phys. A: Math. Gen.* **34** (2001) 5863–874.

23. W.A. Majewski, On non-completely positive quantum dynamical maps on spin chains, *J. Phys. A: Math. Gen.* **40** (2007) 11539–1545.

24. W.A. Majewski, On positive decomposable maps. *Rep. Math. Phys.* **59** (2007), 289–298.

25. F.M. Mukhamedov, On ergodic properties of discrete quadratic dynamical system on C^*-algebras. *Method of Funct. Anal. and Topology*, **7**(2001), No.1, 63–75.

26. F.M. Mukhamedov, On decomposition of quantum quadratic stochastic processes into layer-Markov processes defined on von Neumann algebras, *Izvestiya Math.* **68**(2004), 1009–1024.

27. F. Mukhamedov, On pure quasi-quantum quadratic operators of $M_2(\mathbb{C})$ II, *Open Sys. & Infor. Dyn.* **22** (2015) 1550024.

28. F. Mukhamedov, On circle preserving quadratic operators, *Bull. Malays. Math. Sci. Soc.* **40** (2017), 765–782.

29. F. Mukhamedov, A. Abduganiev, On Kadison-Schwarz type quantum quadratic operators on $M_2(C)$, *Abst. Appl. Anal.* **2013**(2013), Article ID 278606, 9 p.

30. F. Mukhamedov, A. Abduganiev, On description of bistochastic Kadison-Schwarz operators on $M_2(\mathbb{C})$, *Open Systems & Infor. Dynam.* **17**(2010), 245–253.

31. F. Mukhamedov, A. Abduganiev, On pure quasi-quantum quadratic operators of $M_2(\mathbb{C})$, *Open Systems & Infor. Dynam.* **20**(2013), 1350018.

32. F. Mukhamedov, A. Abduganiev, On bistochastic Kadison-Schwarz operators on $M_2(\mathbb{C})$, *Jour. Phys.: Conf. Ser.* **435** (2013), 012018.

33. F. Mukhamedov, H. Akin, S. Temir, A. Abduganiev, On quantum quadrtic operators on $M_2(\mathbb{C})$ and their dynamics, *Jour. Math. Anal. Appl.* **376**(2011) 641–655.

34. M.A. Nielsen, I.L. Chuang, *Quantum Computation and Quantum Information*, Cambridge Univ. Press, Cambridge, 2000.

35. M. Ohya, D. Petz, *Quantum Entropy and Its Use*, Springer, Berlin 1993.

36. M. Ohya, I. Volovich, *Mathematical foundations of quantum information and computation and its applications to nano- and bio-systems*, Springer, New York, 2011.

37. V. Paulsen, *Completely Bounded Maps and Operator Algebras*, Cambridge University Press, 2003.

38. Robertson A.G., Schwarz inequalities and the decomposition of positive maps on C^*-algebras *Math. Proc. Camb. Philos. Soc.* **94**(1983), 291–296.

39. A. G. Robertson, A Korovkin theorem for Schwarz maps on C^*-algebras, *Math. Z.* **156**(1977), 205–206.

40. M.B. Ruskai, S. Szarek, E. Werner, An analysis of completely positive trace-preserving maps on M_2, *Lin. Alg. Appl.* **347** (2002) 159–187.

41. E. Stormer, *Positive linear maps of operator algebras.* Springer, 2013.

Quantum Bio-Informatics VI
© 2020 World Scientific Publishing Co. Pte. Ltd.
pp. 113–128

A CONSTRUCTION OF DYNAMICAL ENTROPY ON CAR ALGEBRAS

KYOUHEI OHMURA* and NOBORU WATANABE[†]

*Department of Information Sciences, Tokyo University of Science,
Noda City, Chiba 278-8510, Japan*
*E-mail: * ohmura.kyouhei@gmail.com, [†] watanabe@is.noda.tus.ac.jp*

Dedicated to Professor Masanori Ohya

The dynamical entropy on von Neumann algebras defined by Accardi, Ohya and Watanabe (AOW entropy) is a natural noncommutative extension of the classical dynamical entropy. On the other hand, the quantum spin lattice systems currently used in quantum computing and communication processes are mathematically described by C*-algebras called CAR algebras. Therefore, in order to obtain the average amount of quantum information of quantum spin systems, it is necessary to define dynamical entropy on CAR algebras. In this paper, we formulate dynamical entropy on CAR algebras based on the construction of the AOW entropy. Moreover, we compute the introduced entropy for a 2×2 matrix algebra case.

Keywords: Quantum Information Theory; Quantum Entropy; Quantum Dynamical Entropy; Quantum Markov Processes; Quantum Statistical Mechanics.

1. Introduction

The classical dynamical entropy for a measure-preserving invertible transformation of a Lebesgue space has important roles in both pure mathematics and classical information theory. In mathematical side, it shows that two classical dynamical systems are isomorphic. In classical information theory, the dynamical entropy gives the average of information amount of an information source [9], [11].

There are several ways to define the dynamical entropy on noncommutative algebras [5], [13], [15]. In [6], using quantum Markov chains on matrix algebras [1], [7], Accardi, Ohya and Watanabe formulated the dynamical entropy on von Neumann algebras. The entropy is called AOW entropy and is a natural noncommutative (or quantum) extension of the classical dynamical entropy. Moreover, due to the simplicity of its formulation, one can easily obtain the average amount of quantum information for quantum dynamical systems using the AOW entropy.

Incidentally, in recent years, quantum spin lattice systems are used in quantum computing and quantum communication processes. Since the observables of spin systems are expressed by the elements of CAR algebras, it is necessary to define dynamical entropy on this algebra to discuss the dynamics and the average information of spin systems strictly.

Therefore, in this paper, we define dynamical entropy on CAR algebras based on the construction method of AOW entropy and the definition of Markov chains on CAR algebras given by Accardi, Fidaleo and Mukhamedov [2].

We organize the paper as follows. In section 2 we recall the definition of the AOW entropy, namely the dynamical entropy through a quantum Markov chain on matrix algebras. In section 3, we briefly review the definitions of the CAR algebra and the Fermion Fock space which gives the algebra. Section 4 is devoted to the notions of Markov states and chains on CAR algebras. In section 5, we formulate dynamical entropy on CAR algebras based on the construction of the AOW entropy, the definition of Markov chains on CAR algebras, and using an Umegaki conditional expectation on CAR algebras. In section 6 we calculate our dynamical entropy for a simple model.

2. AOW Entropy

In this section, we construct dynamical entropy on von Neumann algebras using quantum Markov chains on matrix algebras and a noncommutative extension of measurable partitions of a metric space.

Let - a Hilbert space: $\mathcal{H}$,
- a von Neumann algebra with an identity operator $1_{\mathcal{A}}$ acting on $\mathcal{H}$: $\mathcal{A}$,
- the set of all normal states on $\mathcal{A}$: $\mathfrak{S}(\mathcal{A})$,
- a *-automorphism on $\mathcal{A}$: θ,
- the set of all $d \times d$ matrices on $\mathbb{C}$: M_d,
- the tensor product of $\mathbb{N}$ copies of M_d: $\otimes_{\mathbb{N}} M_d$.

Then the triplet $(\mathcal{A}, \mathfrak{S}(\mathcal{A}), \theta)$ describes the dynamics of a quantum system. Furthermore, let $\gamma := \{\gamma_j\}$ be a finite orthogonal partition of $1_{\mathcal{A}} \in \mathcal{A}$, i.e.

$$\sum_j \gamma_j = 1_{\mathcal{A}} \quad , \quad \gamma_i \gamma_j = \delta_{ij} \gamma_j,$$

and E_e be a completely positive map from $M_d \otimes \mathcal{A}$ to $\mathcal{A}$:

$$E_e\left(\sum_{i,j} e_{ij} \otimes A_{ij}\right) = \sum_i A_{ii}, \qquad (1)$$

with the matrix unit $e_{ij} \in M_d$. Then a transition expectation [1], [3] $\mathcal{E}_{\gamma,\theta}$: $M_d \otimes \mathcal{A} \to \mathcal{A}$ with respect to θ is given by

$$\mathcal{E}_{\gamma,\theta}(a \otimes b) := \theta \circ E_e(p^*_{\gamma,e}(a \otimes b)p_{\gamma,e}) \quad , \quad a \otimes b \in M_d \otimes \mathcal{A}, \qquad (2)$$

where $p_{\gamma,e} := \sum_j e_{jj} \otimes \gamma_j$. In the above notations, a *quantum Markov chain* on $\otimes_{\mathbb{N}} M_d$ is defined by

$$\psi := \{\varphi, \mathcal{E}_{\gamma,\theta}\} \in \mathfrak{S}(\otimes^{\mathbb{N}} M_d), \qquad (3)$$

where φ is called the *initial distribution* of ψ. Then $\psi = \{\varphi, \mathcal{E}_{\gamma,\theta}\}$ is given by

$$\psi(j_1(a_1)j_2(a_2)\cdots j_n(a_n))$$
$$= \varphi(\mathcal{E}_{\gamma,\theta}(a_1 \otimes \mathcal{E}_{\gamma,\theta}(a_2 \otimes \cdots \otimes \mathcal{E}_{\gamma,\theta}(a_n \otimes 1_{\mathcal{A}})\cdots))), \qquad (4)$$

$$n \in \mathbb{N}, a_i \in M_d, \qquad (5)$$

where j_k is an embedding from $a \in M_d$ into the k-th factor of $\otimes_{\mathbb{N}} M_d$, i.e.

$$j_k(a) := 1 \otimes \cdots \otimes 1 \otimes \overset{k-th}{a} \otimes 1 \otimes \cdots \qquad (6)$$

Let φ be a stationary state. Then there exists unique density operator ρ such that

$$\varphi(A) = \mathrm{Tr}\rho A \quad , \quad \forall A \in \mathcal{A}.$$

For any $a_1 \otimes \cdots \otimes a_n \otimes 1_{\mathcal{A}} \in M_d \otimes \cdots \otimes M_d \otimes \mathcal{A}$, we have

$$\psi(j_1(a_1)j_2(a_2)\cdots j_n(a_n)) = \varphi(\mathcal{E}_{\gamma,\theta}(a_1 \otimes \mathcal{E}_{\gamma,\theta}(a_2 \otimes \cdots \otimes \mathcal{E}_{\gamma,\theta}(a_n \otimes 1_{\mathcal{A}}\cdots))))$$

$$= \mathrm{Tr}_{(\otimes^n_1 M_d)\otimes\mathcal{A}} \sum_{i_1}{}^{`}\cdots\sum_{i_{n-1}}\sum_{i_n} e_{i_1 i_1} \otimes \cdots \otimes e_{i_{n-1}i_{n-1}} \otimes e_{i_n i_n}$$
$$\otimes \gamma_{i_n}\theta^*(\gamma_{i_{n-1}}\cdots\theta^*(\gamma_{i_1}\rho\gamma_{i_1})\cdots\gamma_{i_{n-1}})\gamma_{i_n}(a_1 \otimes \cdots \otimes a_{n-1} \otimes a_n \otimes 1_{\mathcal{A}})$$
$$= \mathrm{Tr}_{(\otimes^n_1 M_d)\otimes\mathcal{A}} P_{[0,n]}(a_1 \otimes \cdots \otimes a_{n-1} \otimes 1_{\mathcal{A}})$$
$$= \mathrm{Tr}_{(\otimes^n_1 M_d)} \sum_{i_1}\cdots\sum_{i_{n-1}}\sum_{i_n}(\mathrm{tr}_{\mathcal{A}}\gamma_{i_n}\theta^*(\gamma_{i_{n-1}}\cdots\theta^*(\gamma_{i_1}\rho\gamma_{i_1})\cdots\gamma_{i_{n-1}})\gamma_{i_n})$$
$$e_{i_1 i_1} \otimes \cdots \otimes e_{i_{n-1}i_{n-1}} \otimes e_{i_n i_n}(a_1 \otimes \cdots \otimes a_{n-1} \otimes a_n)$$
$$= \mathrm{Tr}_{(\otimes^n_1 M_d)} \rho_n(a_1 \otimes \cdots \otimes a_{n-1} \otimes a_n),$$

where

$$\rho_{[0,n]} := \sum_{i_1}\cdots\sum_{i_{n-1}}\sum_{i_n} e_{i_1 i_1} \otimes \cdots \otimes e_{i_{n-1}i_{n-1}} \otimes e_{i_n i_n}$$
$$\otimes \gamma_{i_n}\theta^*(\gamma_{i_n-1}\cdots\theta^*(\gamma_{i_1}\rho\gamma_{i_1})\cdots\gamma_{i_{n-1}})\gamma_{i_n},$$
$$\rho_n := \mathrm{Tr}_{\mathcal{A}}\rho_{[0,n]}.$$

Hence we obtain the density operator ρ_n of φ_n on $\otimes_1^n M_d$. Denoting

$$\Gamma_{i_n\cdots i_1} := \theta^{n-1}(\gamma_{i_n})\cdots\theta(\gamma_{i_2})\gamma_{i_1} \tag{7}$$

and

$$P_{i_1,\cdots i_n} := \mathrm{Tr}_{\mathcal{A}}\Gamma_{i_n\cdots i_1}\rho\Gamma^*_{i_n\cdots i_1} = \mathrm{Tr}_{\mathcal{A}}|\Gamma_{i_n\cdots i_1}|^2\rho, \tag{8}$$

Accardi, Ohya and Watanabe defined the entropy with respect to γ,θ and n as

$$S_n(\gamma,\theta) := -\mathrm{Tr}\rho_n\log\rho_n = -\sum_{i_1,\cdots,i_n} P_{i_1,\cdots i_n}\log P_{i_1,\cdots i_n}. \tag{9}$$

Definition 2.1. The dynamical entropy through a quantum Markov chain with respect to γ and θ is given by

$$\tilde{S}(\gamma;\theta) := \limsup_{n\to\infty} \frac{1}{n}S_n(\gamma,\theta)$$
$$= \limsup_{n\to\infty} \frac{1}{n}\left(-\sum_{i_1\cdots i_n} P_{i_1,\cdots,i_n}\log P_{i_1,\cdots,i_n}\right). \tag{10}$$

The above entropy is called *AOW entropy*.

Remark 2.1. Then $P_{i_1,\cdots,i_n}$ (8) is the time ordered correlation kernel [3], [4] over $\mathcal{A}$.

Remark 2.2. The AOW entropy was extended to dynamical mutual entropy and is used to study quantum communication processes [14], [17].

3. CAR Algebras

The CAR algebra is a C^*-algebra generated by the observables of Fermion systems. In this section, we recall the basic mathematical definitions and physical backgrounds of this algebra [10].

3.1. *Fermion Fock Spaces*

First, we fix $N \in \mathbb{N}$. Let $\mathcal{H}$ be a Hilbert space of the state of 1-particle. Then the Hilbert space of states of many body systems is give by

$$\mathcal{H}_N := \bigotimes_N \mathcal{H}. \tag{11}$$

Then $\mathcal{H}_N$ is the Hilbert space that N-particles can be distinguished from each other. Moreover, $\mathcal{P}_N$ denotes a N-dimensional group of permutations, i.e.

$$\mathcal{P}_N := \{\sigma ; \{1, \cdots, N\} \to \{1, \cdots, N\} , \sigma \text{ is injective}\}. \tag{12}$$

If σ is even (resp. odd), we put $\mathrm{sgn}(\sigma) = 1$ (resp. $\mathrm{sgn}(\sigma) = -1$) where sgn is sign of σ.

For each $\sigma \in \mathcal{P}_N$, there exists a unique unitary operator P_σ on $\mathcal{H}_N$ satisfying

$$P_\sigma(x_1 \otimes \cdots \otimes x_N) = x_{\sigma(1)} \otimes \cdots \otimes x_{\sigma(N)} \quad , \quad x_j \in \mathcal{H} , \ j \in \{1, \cdots, N\} \tag{13}$$

and

$$P_\sigma P_\tau = P_{\tau\sigma} \quad , \quad \sigma , \tau \in \mathcal{P}_N. \tag{14}$$

P_σ is called a *permutation operator* with respect to a permutation σ. Using P_σ, we can classify the vectors of $\mathcal{H}_N$.

Definition 3.1. For any $\sigma \in \mathcal{P}_N$,

(1) if $P_\sigma x = x$ holds, x is called *symmetric*.
(2) if $P_\sigma x = \mathrm{sgn}(\sigma)x$ holds, x is called *anti-symmetric*.

Put

$$\bigwedge^N(\mathcal{H}) := \{x \in \mathcal{H}_N ; x \text{ is anti-symmetric}\}. \tag{15}$$

The elements of $\bigwedge^N(\mathcal{H})$ are called *Fermions*.

Remark 3.1. For any $x_1, \cdots, x_N \in \mathcal{H}$,

$$x_1 \bigwedge x_2 \bigwedge \cdots \bigwedge x_N := \frac{1}{\sqrt{N!}} \sum_{\sigma \in \mathcal{P}_N} \mathrm{sgn}(\sigma) x_{\sigma(1)} \otimes \cdots \otimes x_{\sigma(N)}$$

is an anti-symmetric vector. For any $i, j \in \{1, \cdots, N\}$ $(i \neq j)$,

$$x_1 \bigwedge \cdots \bigwedge x_i \bigwedge \cdots \bigwedge x_j \bigwedge \cdots \bigwedge x_N$$
$$= -x_1 \bigwedge \cdots \bigwedge x_j \bigwedge \cdots \bigwedge x_i \bigwedge \cdots \bigwedge x_N$$

holds. Especially, if $x_i = x_j$, one can see that

$$x_1 \bigwedge x_2 \bigwedge \cdots \bigwedge x_N = 0. \tag{16}$$

(16) implies that two fermions can not exist in the same 1-particle state. The property is called the *Pauli exclusion principle*. We will see the algebraic representation of this principle below.

In general, the number of particles in many body systems can change. Therefore one has to consider the generalized Hilbert space of $\mathcal{H}_N$:

$$\mathcal{F}(\mathcal{H}) := \bigoplus_{N=0}^{\infty} \mathcal{H}_N \tag{17}$$

$$= \left\{ x = \{x^{(N)}\}_{N=0}^{\infty} \ ; \ x^{(N)} \in \mathcal{H}_N, N \geq 0, \ \sum_{N=0}^{\infty} \|x^{(N)}\|_{\mathcal{H}_N}^2 < \infty \right\}$$

where $\mathcal{H}_0 = \mathbb{C}$. Then the scalar product of $\mathcal{F}(\mathcal{H})$ is defined by

$$\langle x, y \rangle := \sum_{N=0}^{\infty} \left\langle x^{(N)}, y^{(N)} \right\rangle_{\mathcal{H}_N}.$$

Hilbert space $\mathcal{F}(\mathcal{H})$ is called a *full Fock space*.

Definition 3.2. Let $\bigwedge^0(\mathcal{H}) := \mathbb{C}$.

$$\mathcal{F}_f(\mathcal{H}) := \bigoplus_{N=0}^{\infty} \bigwedge^N(\mathcal{H}) \tag{18}$$

$$= \left\{ x = \{x^{(N)}\}_{N=0}^{\infty} \ ; \ x^{(N)} \in \bigwedge^N(\mathcal{H}), \ N \geq 0, \ \sum_{N=0}^{\infty} \|x^{(N)}\|^2 < \infty \right\}$$

is called a *Fermion Fock space*.

$\mathcal{F}_f(\mathcal{H})$ is a closed subspace of the full Fock space $\mathcal{F}(\mathcal{H})$.

Definition 3.3. A subspace of $\mathcal{F}_f(\mathcal{H})$:

$$\mathcal{F}_{f,0}(\mathcal{H}) := \{x \in \mathcal{F}_f(\mathcal{H}) \ ; \ \text{If there exists } N_0 \text{ such that } N \geq N_0, \ x^{(N)} = 0\} \tag{19}$$

is called a *finite particles subspace* and is dense in $\mathcal{F}_f(\mathcal{H})$.

3.2. *CAR Algebras*

Let $\mathcal{D}$ be a dense subspace of $\mathcal{H}$ and let

$$\mathcal{F}_{f,fin}(\mathcal{D}) := \left\{ x \in \mathcal{F}_f(\mathcal{H}) \; ; \; x^{(N)} \in A_N(\widehat{\otimes}^N \mathcal{D}), \; N \geq 0, \right. \tag{20}$$

$$\left. \text{If there exists } N_0 \text{ such that } N \geq N_0, \; x^{(N)} = 0 \right\} \tag{21}$$

where $A_N := \frac{1}{N!}\sum_{\sigma \in \mathcal{P}_N} P_\sigma$. Then $\mathcal{F}_{f,fin}(\mathcal{D})$ is called a *finite particles subspace on* $\mathcal{D}$ and is dense in $\mathcal{F}_f(\mathcal{H})$.

For each $x \in \mathcal{H}$, we define the operator $a_+(f)$ on $\mathcal{F}_f(\mathcal{H})$ as follows:

$$\text{Dom}(a_+) := \left\{ x \in \mathcal{F}_f(\mathcal{H}) \; ; \; \sum_{N=0}^{\infty} N \|A_N(f \otimes x^{(N-1)})\|^2 < \infty \right\}, \tag{22}$$

$$(a_+(f)x)^{(N)} := \sqrt{N} A_N(f \otimes x^{(N-1)}) \; (N \geq 1) \quad , \quad (a_+(f)x)^{(0)} := 0. \tag{23}$$

One can know that $b_+(f)$ is closed. Since $\mathcal{F}_{f,fin}(\mathcal{D}) \subset \text{Dom}(a_+(f))$ holds, $a_+(f)$ is densely defined. Therefore the adjoint operator

$$a(f) := a_+(f)^* \tag{24}$$

exists. Furthermore, since $a_+(f)$ is closed,

$$a(f)^* := a_+(f). \tag{25}$$

$a(f)^*$ is called a *Fermion creation operator* and $a(f)$ is called a *Fermion annihilation operator* respectively.

Then the following proposition is satisfied. Let Ω be a vacuum state.

Proposition 3.1. *For any* $f_j \in \mathcal{H}$ $(j = 1, \cdots, n)$,

$$(a(f_1)^* a(f_2)^* \cdots a(f_n)^* \Omega)^{(n)} = \sqrt{n!} A_n(f_1 \otimes \cdots \otimes f_n), \tag{26}$$

$$(a(f_1)^* a(f_2)^* \cdots a(f_n)^* \Omega)^{(N)} = 0 \quad , \quad N \neq 0. \tag{27}$$

The Fermion operators satisfy the following algebraic relations.

Theorem 3.1. *For each* $f \in \mathcal{H}$, $a(f), a(f)^* \in \mathbf{B}(\mathcal{F}_f(\mathcal{H}))$ *and the following relation hold:*

$$\{a(f), a(g)^*\} = \langle f, g \rangle \quad , \quad f, g \in \mathcal{H}, \tag{28}$$

$$\{a(f), a(g)\} = 0 \quad , \quad \{a(f)^*, a(g)^*\} = 0 \quad , \quad f, g \in \mathcal{H}, \tag{29}$$

where $\{\cdot, \cdot\}$ *is the anti commutator* $\{A, B\} := AB + BA$.

In (32), if $f = g$, there hold

$$a(f)^2 = 0 \quad , \quad (a(f)^*)^2 = 0. \tag{30}$$

These are the algebraic representations of the Pauli exclusion principle (16).

Now we state the definition of the CAR algebra.

Definition 3.4. A C^*-algebra generated by bounded operators $\{a(f), a(f)^* ; f \in \mathcal{H}\}$ is called a CAR algebra.

The CAR algebra is a C^*-algebra of observables of Fermion systems.

4. Markov States and Chains on CAR Algebras

In section 3, we showed that the CAR algebra describes the Fermi systems. In this section, we construct Markov chains on CAR algebras following [2]. Since the Markov chain is a stochastic process, it is natural to consider on $\mathbb{Z}_+ := \mathbb{N} \cup \{0\}$ as follows.

Let $\mathcal{A}_{\mathbb{Z}_+}$ be the CAR algebra generated by $\{a_i, a_i^* ; i \in \mathbb{Z}_+\}$. Namely, $\{a_i, a_i^* ; i \in \mathbb{Z}_+\}$ satisfy:

$$(a_i)^* = a_i^* \quad , \quad \{a_i^*, a_j\} = \delta_{ij} 1, \tag{31}$$

$$\{a_i, a_j\} = \{a_i^*, a_j^*\} = 0 \quad , \quad i, j \in \mathbb{Z}_+, \tag{32}$$

where $\{\cdot, \cdot\}$ is the anti-commutator. For any $I \subset \mathbb{Z}_+$, $\mathcal{A}_I$ denotes the subalgebra generated by $\{a_i, a_i^* ; i \in I\}$. In particular,

$$\mathcal{A}_{n]} := \mathcal{A}_0 \bigvee \mathcal{A}_1 \bigvee \cdots \bigvee \mathcal{A}_n.$$

Remark 4.1. In general, the *-algebra generated by a subset $\mathcal{M} \subset \mathcal{A}$ is given by

$$\text{l.i.h.}(\{M_1 \cdots M_n ; n \in \mathbb{N}, M_j \in \mathcal{M} \cup \mathcal{M}^*, j = 1, \cdots, n\})$$

$$= \left\{ \sum_{n=1}^{N} \lambda_n M_1^{(n)} \cdots M_m^{(n)} ; N \in \mathbb{N}, \lambda \in \mathbb{C} \right\}. \tag{33}$$

Especially, in CAR case, the *-algebra generated by $\{a_i, a_i^* ; i \in \{k\}\}$ is written as

$$\mathcal{A}_k := \left\{ \sum_{n=1}^{N} \lambda_n A_{i_1}^{(n)} \cdots A_{i_m}^{(n)} ; A_{i_k}^{(n)} \in \{a_k, a_k^*\}, N \in \mathbb{N}, \lambda_n \in \mathbb{C} \right\}. \tag{34}$$

Due to the relations (31)-(32), $\mathcal{A}_k$ becomes

$$\mathcal{A}_k = \{\lambda_1 a_k + \lambda_2 a_k^* + \lambda_3 a_k^* a_k + \lambda_4 a_k a_k^* \; ; \; \lambda_n \in \mathbb{C}\}. \tag{35}$$

Therefore, we only consider the elements $\{a_k, a_k^*, a_k^* a_k, a_k a_k^*\}$ in $I = \{k\}$, $k \in \mathbb{Z}_+$ case.

Put $\mathcal{A} := \mathcal{A}_{\mathbb{Z}_+}$.

Definition 4.1. We denote Θ_I as the unique automorphism on $\mathcal{A}$ which satisfy

$$\Theta_I(a_i) = -a_i \quad , \quad \Theta_I(a_i^*) = -a_i^* \quad , \quad (i \in I) \tag{36}$$

$$\Theta_I(a_i) = a_i \quad , \quad \Theta_I(a_i^*) = a_i^* \quad , \quad (i \in I^c) \tag{37}$$

Θ_I is called a *parity automorphism* on $\mathcal{A}$.

Let $\Theta := \Theta_{\mathbb{Z}_+}$. Then Θ induces even and odd parts of $\mathcal{A}$:

$$\mathcal{A}_+ := \{a \in \mathcal{A} \; ; \; \Theta(a) = a\} \quad , \quad \mathcal{A}_- := \{a \in \mathcal{A} \; ; \; \Theta(a) = -a\}. \tag{38}$$

Definition 4.2. Let $\mathcal{A}$, $\mathcal{B}$ be CAR algebras and E be a map from $\mathcal{A}$ to $\mathcal{B}$. If

$$E \circ \Theta = E \tag{39}$$

holds, E is called *even*.

If E is even,

$$E(a) = E(\Theta(a)) = -E(a) = 0 \tag{40}$$

holds for each $a \in \mathcal{A}$.

Definition 4.3. Let φ be a state on $\mathcal{A}$. If there exists a quasi-conditional expectation E_n w.r.t. $\mathcal{A}_{n-1]} \subset \mathcal{A}_{n]} \subset \mathcal{A}_{n+1]}$ which satisfy

$$\varphi_{n]} \circ E_n = \varphi_{n+1]}, \tag{41}$$

$$E_n(\mathcal{A}_{[n,n+1]}) \subset \mathcal{A}_n, \tag{42}$$

φ is called a *Markov state*.

Let $\{\mathcal{E}_n\}_{n\in\mathbb{N}}$ be a sequence of completely positive identity-preserving maps satisfying for each $n \in \mathbb{N}$:

$$\mathcal{E}_n : \mathcal{A}_{[0,n+1]} \to \mathcal{A}_{[0,n]} \tag{43}$$

$$\mathcal{E}_{n+1}|_{[0,n]} = 1_{[0,n]}, \tag{44}$$

$$\mathcal{E}_{n+1} \circ \alpha|_{[0,n+1]} = \alpha|_{[0,n]} \circ \mathcal{E}_n, \tag{45}$$

where α is the one-step right shift. One can see that (45) gives the following diagram.

$$\begin{array}{ccc} \mathcal{A}_{[0,n+1]} & \xrightarrow{\alpha|_{[0,n+1]}} & \mathcal{A}_{[0,n+2]} \\ \mathcal{E}_n \downarrow & \xrightarrow{} & \downarrow \mathcal{E}_{n+1} \\ \mathcal{A}_{[0,n]} & \xrightarrow{\alpha|_{[0,n]}} & \mathcal{A}_{[0,n+1]} \end{array} \tag{46}$$

Let ρ be a density operator on $\mathcal{A}_0$ which satisfies the stationarity in the sense of

$$\rho = \rho \circ \mathcal{E}_0 \circ \alpha|_0. \tag{47}$$

In the above notations one obtain the sequence of states $\{\varphi_n\}_{n\in\mathbb{N}}$:

$$\varphi_n := \mathrm{Tr}\rho \circ \mathcal{E}_0 \circ \cdots \circ \mathcal{E}_n. \tag{48}$$

Due to (44), (45) and (47), one can extend φ_n to a shift-invariant state φ on $\mathcal{A}_{\mathbb{N}}$. The shift-invariant state is called the *quantum Markov chain* generated by $(\rho, \{\mathcal{E}_n\}_{n\in\mathbb{N}})$.

5. A Construction of Dynamical Entropy on CAR Algebras

In this section, giving an Umegaki conditional expectation [8], [16], we formulate dynamical entropy on CAR algebras.

Let $\mathcal{A}_0$ be a CAR algebra generated by $\{a_i, a_i^* ; i \in \{0\}\}$ and φ_0 a faithful normal state on $\mathcal{A}_0$. $\mathcal{E}_{\gamma,\theta}$ denotes an Umegaki conditional expectation from $\mathcal{A}_{\{0,n\}}$ to $\mathcal{A}_0$ as

$$\mathcal{E}_{\gamma,\theta} := \frac{1}{2}\left(\mathrm{Tr}_n\left(\Theta^n \otimes \sum_{i=1}^{2} \theta^{n-1}(\gamma_i)^*(1_0)\theta^{n-1}(\gamma_i) \right) \right) \tag{49}$$

where 1_0 is the identity of $\mathcal{A}_0$, θ is a *-automorphism on $\mathcal{A}_0$, and $\sum_{i=1}\gamma_i = 1_0$ is an operator partition of 1_0. Since $\theta(1_0) = 1_0$, $\sum\theta(\gamma_i)$ is again operator partition of 1_0.

Now we construct Markov chain with 1-time evolution:

$$\varphi_0(\mathcal{E}_{\gamma,\theta}(A_1 \otimes 1_0)) = \mathrm{Tr}_0 \rho_0(\mathcal{E}_{\gamma,\theta}(A_1 \otimes 1_0))$$

$$= \mathrm{Tr}_0 \rho_0 \left(\frac{1}{2} \left(\mathrm{Tr}_1 \left(\Theta^1 \otimes \sum_i \theta(\gamma_i)^* \theta(\gamma_i) \right) (A_1 \otimes 1_0) \right) \right)$$

$$= \frac{1}{2} \mathrm{Tr}_0 \rho_0 \left(\mathrm{Tr}_1 \left(A_{1,+} \otimes \sum_i \theta(\gamma_i)^* \theta(\gamma_i) \right) \right)$$

$$= \mathrm{Tr}_{[0,1]} \left(\frac{1}{2} A_{1,+} \otimes \sum_i \rho_0 \theta(\gamma_i)^* \theta(\gamma_i) \right).$$

Since

$$A_{k,+} = a_k a_k^* + a_k^* a_k = 1 \quad , \quad \sum_i \theta(\gamma_i)^* \theta(\gamma_i) = \theta(\sum_i \gamma_i) = 1,$$

we have

$$\mathrm{Tr}_k \frac{1}{2} A_{k,+} = 1 \quad , \quad \mathrm{Tr}_0 \rho_0 \theta(\gamma_i)^* \theta(\gamma_i) = 1$$

respectively. Therefore, we obtain a state through a Markov chain on $\mathcal{A}_{[0,1]}$ as

$$\rho_{[0,1]} := \frac{1}{2} A_{1,+} \otimes \sum_i \rho_0 \theta(\gamma_i)^* \theta(\gamma_i). \tag{50}$$

Moreover, we consider the case of n-time evolution:

$$\varphi_0(\mathcal{E}_{\gamma,\theta}(A_1 \otimes \mathcal{E}_{\gamma,\theta}(A_2 \otimes \cdots \mathcal{E}_{\gamma,\theta}(A_{n-1} \otimes \mathcal{E}_{\gamma,\theta}(A_n \otimes 1_0)) \cdots)))$$

$$= \varphi_0 \left(\mathcal{E}_{\gamma,\theta} \left(A_1 \otimes \mathcal{E}_{\gamma,\theta} \left(A_2 \otimes \cdots \mathrm{Tr}_n \left(\frac{1}{2} \left(\Theta^n \otimes \sum_{i_n} \theta(\gamma_{i_n})^* \theta(\gamma_{i_n}) \right. \right. \right. \right. \right.$$
$$\left. \left. \left. \left. \left. (A_n \otimes 1_0) \right) \right) \right) \cdots \right) \right)$$

$$= \varphi_0 \left(\mathcal{E}_{\gamma,\theta} \left(A_1 \otimes \mathcal{E}_{\gamma,\theta} \left(A_2 \otimes \cdots \mathrm{Tr}_n \left(\frac{1}{2} \left(A_{n,+} \otimes \sum_{i_n} \theta(\gamma_{i_n})^* \theta(\gamma_{i_n}) \right) \right) \cdots \right) \right) \right)$$

$$= \frac{1}{2} \mathrm{Tr}_n(A_{n,+})$$
$$\cdot \varphi_0 \left(\mathcal{E}_{\gamma,\theta} \left(A_1 \otimes \cdots \mathcal{E}_{\gamma,\theta} \left(A_{n-1} \otimes \sum_{i_n} \theta(\gamma_{i_n})^* \theta(\gamma_{i_n}) \right) \cdots \right) \right)$$

$$= \frac{1}{2}\mathrm{Tr}_n(A_{n,+}) \cdot$$

$$\cdot \varphi_0\Bigg(\mathcal{E}_{\gamma,\theta}\Bigg(A_1 \otimes \cdots \mathrm{Tr}_{n-1}\Bigg(\frac{1}{2}\Bigg(A_{n-1,+} \otimes$$

$$\sum_{i_n,i_{n-1}} \theta(\gamma_{i_{n-1}}^* \theta(\gamma_{i_n}^* \gamma_{i_n})\gamma_{i_{n-1}})\Bigg)\Bigg)\cdots\Bigg)\Bigg)$$

$$= \frac{1}{2^2}\mathrm{Tr}_{n,n-1}(A_{n,+} \otimes A_{n-1,+})$$

$$\cdot \varphi_0\Bigg(\mathcal{E}_{\gamma,\theta}\Bigg(\cdots \mathcal{E}_{\gamma,\theta}\Bigg(A_{n-2} \otimes \sum_{i_n,i_{n-1}} \theta(\gamma_{i_{n-1}}^* \theta(\gamma_{i_n}^* \gamma_{i_n})\gamma_{i_{n-1}})\Bigg)\cdots\Bigg)\Bigg)$$

$$\vdots$$

$$= \frac{1}{2^n}\mathrm{Tr}_{\{n,\cdots,1\}}(A_{n,+} \otimes A_{n-1,+} \otimes \cdots \otimes A_{1,+})$$

$$\cdot \varphi_0(\sum_{i_n,\cdots,i_1} \theta(\gamma_{i_1}^* \cdots \theta(\gamma_{i_n}^* \gamma_{i_n}) \cdots \gamma_{i_1}))$$

$$= \frac{1}{2^n}\mathrm{Tr}_{\{n,\cdots,1\}}(A_{n,+} \otimes A_{n-1,+} \otimes \cdots \otimes A_{1,+})$$

$$\cdot \mathrm{Tr}_0\rho_0(\sum_{i_n,\cdots,i_1} \theta(\gamma_{i_1}^* \cdots \theta(\gamma_{i_n}^* \gamma_{i_n}) \cdots \gamma_{i_1}))$$

$$= \frac{1}{2^n}\mathrm{Tr}_{\{0,\cdots,n\}}\Bigg(A_{n,+} \otimes A_{n-1,+} \otimes \cdots \otimes A_{1,+} \otimes$$

$$\sum_{i_1,\cdots,i_n} \rho\theta(\gamma_{i_1})^* \theta^2(\gamma_{i_2})^* \cdots \theta^n(\gamma_{i_n}^* \gamma_{i_n}) \cdots \theta^2(\gamma_{i_2})\theta(\gamma_{i_1})\Bigg).$$

Hence a state thorough a Markov chain on $\mathcal{A}_{[0,n]}$ is given by

$$\rho_{[0,n]} := \frac{1}{2^n}\Bigg(A_{n,+} \otimes A_{n-1,+} \otimes \cdots \otimes A_{1,+} \otimes \tag{51}$$

$$\sum_{i_1,\cdots,i_n} \theta^n(\gamma_{i_n}) \cdots \theta^2(\gamma_{i_2})\theta(\gamma_{i_1})\rho\theta(\gamma_{i_1})^* \theta^2(\gamma_{i_2})^* \cdots \theta^n(\gamma_{i_n})^*\Bigg). \tag{52}$$

Therefore, we get an n-time developed state on $\mathcal{A}_{[1,n]}$ as

$$\rho_n := \frac{1}{2^n}\sum_{i_1,\cdots,i_n} \mathrm{Tr}_0 \theta^n(\gamma_{i_n}) \cdots \theta^2(\gamma_{i_2})\theta(\gamma_{i_1})\rho\theta(\gamma_{i_1})^* \theta^2(\gamma_{i_2})^* \cdots \theta^n(\gamma_{i_n})^*$$

$$\times A_{n,+} \otimes A_{n-1,+} \otimes \cdots \otimes A_{1,+} \tag{53}$$

Now we focus on the constant of ρ_n. Denote

$$P_{i_1,\cdots,i_n} := \mathrm{Tr}_0 \theta^n(\gamma_{i_n}) \cdots \theta^2(\gamma_{i_2})\theta(\gamma_{i_1})\rho\theta(\gamma_{i_1})^*\theta^2(\gamma_{i_2})^* \cdots \theta^n(\gamma_{i_n})^*. \tag{54}$$

Obviously, then there holds

$$\sum P_{i_1,\cdots,i_n} = 1.$$

Hence, in the above notations, we can define dynamical entropy on the CAR algebra as following.

Definition 5.1. For a quadruple $(\mathcal{A}, \varphi_0, \gamma, \theta)$, the *dynamical entropy* on the CAR algebra is given by

$$h_{\varphi_0}(\theta) := \sup_\gamma \left\{ -\limsup_{n\to\infty} \frac{1}{n} \sum_{i_1,\cdots,i_n} P_{i_1,\cdots,i_n} \log P_{i_1,\cdots,i_n} \right\}. \tag{55}$$

6. Model Computation

In this section, we calculate the introduced dynamical entropy (55) for a simple model.

Let

$$a_0^* := \begin{pmatrix} 0 & 1 \\ 0 & 0 \end{pmatrix} \quad, \quad a_0 := \begin{pmatrix} 0 & 0 \\ 1 & 0 \end{pmatrix}. \tag{56}$$

Since

$$a_0^* a_0 = \begin{pmatrix} 1 & 0 \\ 0 & 0 \end{pmatrix} \quad, \quad a_0 a_0^* = \begin{pmatrix} 0 & 0 \\ 0 & 1 \end{pmatrix},$$

$\mathcal{A}_0 := M(2, \mathbb{C})$ is generated by $\{a_0^*, a_0\}$. According to

$$a_0^* a_0 + a_0 a_0^* = 1 \quad, \quad (a_0^* a_0)^* = a_0^* a_0,$$

we put:
An operator partition of the identity $1 \in \mathcal{A}_0$:

$$\gamma_1 := a_0^* a_0 \quad, \quad \gamma_2 := a_0 a_0^*, \tag{57}$$

a *-automorphism:

$$\theta(a) := UaU^* \equiv e^{ia_0^* a_0} a e^{-ia_0^* a_0} =$$

$$= \begin{pmatrix} e^i & 0 \\ 0 & 1 \end{pmatrix} a \begin{pmatrix} e^{-i} & 0 \\ 0 & 1 \end{pmatrix} = (e^i a_0^* a_0 + a_0 a_0^*) a (e^{-i} a_0^* a_0 + a_0 a_0^*) \quad, \quad a \in \mathcal{A}_0. \tag{58}$$

126

Moreover, we denote a density matrix by

$$\rho := \begin{pmatrix} \lambda & 0 \\ 0 & 1-\lambda \end{pmatrix} = \lambda a_0^* a_0 + (1-\lambda) a_0 a_0^* \quad , \quad 0 \le \lambda \le 1. \tag{59}$$

Due to

$$\theta^n(\gamma_{i_n})\theta^{n-1}(\gamma_{i_{n-1}})\cdots\theta(\gamma_{i_1})$$

$$= U^n \gamma_{i_n} U^{*n} U^{n-1} \gamma_{i_{n-1}} U^{*n-1} \cdots U \gamma_{i_1} U^*$$

$$= U^n \gamma_{i_n} U^* \gamma_{i_{n-1}} U^* \cdots U^* \gamma_{i_1} U^*,$$

(54) becomes

$$P_{i_1,\cdots,i_n} = \mathrm{Tr} U^n \gamma_{i_n} U^* \gamma_{i_{n-1}} U^* \cdots U^* \gamma_{i_1} U^* \rho U \gamma_{i_1} U \cdots U \gamma_{i_{n-1}} U \gamma_{i_n} U^{*n}$$

$$= \mathrm{Tr} \gamma_{i_n} U^* \gamma_{i_{n-1}} U^* \cdots U^* \gamma_{i_1} U^* \rho U \gamma_{i_1} U \cdots U \gamma_{i_{n-1}} U \gamma_{i_n}.$$

Now we mention the following result.

Lemma 6.1. *Under the above notations,*

$$P_{1,1,\cdots,1} = \lambda \ , \ P_{1,2,\cdots,1} = \cdots = P_{2,2,\cdots,1} = 0 \ , \ P_{2,2,\cdots,2} = 1 - \lambda. \tag{60}$$

Proof. According to the anti-commutation relation (32),

$$U\gamma_1 = (e^i a_0^* a_0 + a_0 a_0^*) a_0^* a_0 = e^i a_0^* a_0.$$

The above equation gives

$$U\gamma_1 U\gamma_1 \cdots \overset{n-th}{U\gamma_1} = e^{ni} a_0^* a_0 \quad , \quad \gamma_1 U^* \gamma_1 \cdots \overset{n-th}{\gamma_1} U^* = e^{-ni} a_0^* a_0.$$

Therefore, if $(i_1, i_2, \cdots, i_n) = (1, 1, \cdots, 1)$, there holds

$$P_{1,1,\cdots,1} = \mathrm{Tr} \gamma_1 U^* \gamma_1 U^* \cdots U^* \gamma_1 U^* \rho U \gamma_1 U \cdots U \gamma_1 U \gamma_1$$

$$= \mathrm{Tr} e^{-ni} a_0^* a_0 (\lambda a_0^* a_0 + (1-\lambda) a_0 a_0^*) e^{ni} a_0^* a_0 = \mathrm{Tr} \lambda a_0^* a_0 = \lambda. \tag{61}$$

Moreover,

$$\gamma_1 U \gamma_2 = a_0^* a_0 (e^i a_0^* a_0 + a_0 a_0^*) a_0 a_0^* = a_0^* a_0 \cdot a_0 a_0^* = 0.$$

This implies that

$$P_{1,2,\cdots,1} = \cdots = P_{2,2,\cdots,1} = 0. \tag{62}$$

Finally, the equation:

$$U\gamma_2 = (e^i a_0^* a_0 + a_0 a_0^*) a_0 a_0^* = a_0 a_0^*$$

induces

$$U\gamma_2 U\gamma_2 \cdots \overset{n-th}{U}\gamma_2 = a_0 a_0^* \quad , \quad \gamma_2 U^* \gamma_2 \cdots \overset{n-th}{\gamma_2 U^*} = a_0 a_0^*.$$

Hence we obtain

$$P_{2,2,\cdots,2} = \mathrm{Tr} a_0 a_0^* (\lambda a_0^* a_0 + (1-\lambda) a_0 a_0^*) a_0 a_0^* = \mathrm{Tr}(1-\lambda) a_0 a_0^* = 1-\lambda. \quad (63)$$

$\square$

From the above lemma, we then get the dynamical entropy (55):

$$h_\rho(\theta) = -\lim_{n\to\infty} \frac{1}{n}((1-\lambda)\log(1-\lambda) + \lambda\log\lambda) = 0. \quad (64)$$

Remark 6.1. In general, the complexities of unitary operators on a unital C^*-algebra are 0 [13]. Therefore, this result tells us that our dynamical entropy is defined correctly.

7. Conclusion

In this paper, based on the construction method of the AOW entropy, we have defined a new type of dynamical entropy on CAR algebras using an Umegaki conditional expectation from $\mathcal{A}_{\{0,n\}}$ to $\mathcal{A}_0$ with a *-automorphisms on a local algebra $\mathcal{A}_0$. Moreover, we have computed the introduced entropy for a 2×2 matrix algebra case.

Incidentally, if one investigate the complexities of lattice translation or Bogoliubov automorphism using dynamical entropy through a Markov chain, we think that it is necessary to specifically give a completely positive identity preserving map from $\mathcal{A}_{[0,n+1]}$ to $\mathcal{A}_{[0,n]}$ associated with a *-automorphism on $\mathcal{A}_{\mathbb{Z}_+}$.

Acknowledgements The authors would like to thank Prof. Accardi for his useful suggestions and warm encouragements.

References

1. L. Accardi: Noncommutative Markov chains, in School of Mathematical Physics, Camerino, 268-295 (1974).
2. L. Accardi, F. Fidaleo, F. Mukhamedov : Markov states and chains on the CAR algebra, Inf. Dymen. Anal. Quantum Probab., Rel. Top. **10**, 165-184 (2007).
3. L. Accardi, A. Frigerio, J. Lewis : Quantum stochastic processes. Publ. RIMS Kyoto Univ. 18, 97-133 (1982).
4. L. Accardi, N. Obata : Foundations of Quantum Probability Theory. Makino Pub. Co., Tokyo (2003).

5. L. Accardi, M. Ohya, N. Watanabe : Note on quantum dynamical entropies, Report on Math. Phys., **38**, 457-469 (1996).

6. L. Accardi, M. Ohya, N. Watanabe : Dynamical entropy through quantum Markov chains, Open Sys. & Information Dyn. **4**, 71-87 (1997).

7. L. Accardi, A. Souissi, S. Gheteb : Quantum Markov Chains: unification approach, arXiv: 1811.00500, 1 Nov (2018).

8. H. Araki, H. Moriya : Equilibrium Statistical Mechanics of Fermion Lattice Systems, Rev. in Math. Phys., **15**, 02, pp. 93-198 (2003).

9. L. Bilingsley : Ergodic Theory and Information, Wiley, New York (1965).

10. O. Bratteli, D. W. Robinson : Opertor Algebras and Quantum Statistical Mechanics II, Springer, New York (1987).

11. A. N. Kolmogorov : Theory of transmission of information. Am. Math. Soc. Transl. Ser. 2 **33**, 291-321 (1963).

12. H. Moriya : Separability condition for composite systems of distinguishable fermions, arXiv : quant-ph/0405166v2, 5 Sep (2005).

13. S. Neshveyev, E. Størmer : Dynamical Entropy in Operator Algebras, Springer, Berlin (2006).

14. K. Ohmura, N. Watanabe : Quantum Dynamical Mutual Entropy Based on AOW Entropy, Open Syst. Inf. Dyn., **26**, 2, pp. 1950009-1 - pp. 1950009-16 (2019).

15. M. Ohya, D. Petz : Quantum Entropy and its Use, Springer, Berlin (1993).

16. H. Umegaki : Conditional expectations in an operator algebra, IV, (entropy and information), Kodai Mathematical Seminar Reports, **14**, 59-85 (1962).

17. N. Watanabe, M. Muto : Note on transmitted complexity for quantum dynamical systems, Philosophical Transactions Royal Society **A 375**, pp. 20160396-1-pp. 20160936-16 (2017)

Quantum Bio-Informatics VI
© 2020 World Scientific Publishing Co. Pte. Ltd.
pp. 129–138

LOCAL GAUGE INVARIANCE AND SYMMETRY BREAKING IN CATEGORICAL QFT

IZUMI OJIMA

RIMS, Kyoto University, Japan
E-mail: ojima@za.ztv.ne.jp

From the viewpoint of "geometry of symmetry breaking", universal roles played by holonomy terms have been found in relation with Elie Cartan's characterization of symmetric spaces: they can be regarded as geometric templates in the physical emergence processes of Macro classical objects from Micro quantum dynamics. In view of the essential roles played by natural transformations here, the logical essence of the emergences can be found in the local gauge invariance, which entails the validity of Maxwell type equations.

1. Introduction

To clarify the close relationship among symmetry breaking, local gauge invariance and Maxwell-type equations, we discuss the following basic points:

1) Quadrality scheme [1,2] as a framework for going back & forth between *Macro* and *Micro* levels of nature:

> *Visible phenomenological Macro data*
> ⇌ *theory of invisible Micro processes*;

2) In algebraic & categorical QFT we show how *local gauge invariance* arises from symmetry breaking;

3) From the viewpoint of broken symmetries & local gauge invariance, basic ingredients of the formalism are reviewed, in which symmetric space structure is found in the sector-classifying space.

2. Basic Concepts: Quadrality Scheme & Micro-Macro Duality based on Sectors

Quadrality Scheme,

	Spec	
States ⇄↑ (Rep) ↓⇄ Alg		
	Dyn	↗

, for describing physical

phenomena is composed of the following four basic ingredients:

Alg(ebra of physical variables)/ **States** (as expectation functionals)/

Spec (as a classifying space of sectors)/ **Dyn**(amics),

forming **Micro-Macro Duality** [3]:

Macro: $\mathcal{A}$	left adjoint F
$\searrow\nwarrow$	
right adjoint E	$\mathcal{X}$: Micro

.

1) Its Micro-Macro boundary is defined in terms of **sectors** and

2) the Macro side is epigenetically due to the **emergence process** of

3) Spec = sector-classifying space from Micro dynamics, to form a categorical adjunction:

$\mathcal{A}(a \leftarrow Fx)$	$\varepsilon_a F(-)$
$\searrow\nwarrow$	
$E(-)\eta_x$	$\mathcal{X}(Ea \leftarrow x)$

,

with unit $\eta : I_{\mathcal{X}} \dot{\to} T := EF$ intertwining $\mathcal{X}$ to monad $T \curvearrowright \mathcal{X}$ as Micro dynamics and with counit $\varepsilon : S := FE \dot{\to} I_{\mathcal{A}}$ to $\mathcal{A}$ from comonad $S \curvearrowright \mathcal{A}$ as dual of monad T.

Micro-Macro duality as categorical adjunction:

	comonad $S = FE$: Spec		
emergence $\nearrow$	Arveson spec $\quad V \uparrow\downarrow I \quad$ spectral subspace		$\searrow$ quantum fields
States $\mathcal{A}$	$\leftrightarrows$ Bimodule of adjoint pair $\leftrightarrows$ $\xleftarrow{F}$ $\xrightarrow{E}$		$\searrow$ Algebra $\mathcal{X}$
	$\uparrow\downarrow$: Gal		$\nearrow$ co-emergence
	Dyn = monad $T = EF$		

3. Sectors & Spec = sector-classifying space

Basic ingredients of the formalism [1,2] are defined as follows:

1) **Sectors** = **pure phases** parametrized by **order parameter**[= central observables $\mathfrak{Z}_{\pi}(\mathcal{X}) = \pi(\mathcal{X})'' \cap \pi(\mathcal{X})'$ commuting with all physical variables $\pi(\mathcal{X})''$ in a generic representation π of algebra $\mathcal{X}$ of physical variables]:

Mathematically, a **sector**(= **pure phase**) $\overset{\text{def}}{=}$ a **quasi-equivalence** *class of factor* states (& representations π_γ) of (C*-)algebra $\mathcal{X}$ of physical variables, as a *minimal unit* of representations characterized by *trivial centre* $\pi_\gamma(\mathcal{X})'' \cap \pi_\gamma(\mathcal{X})' =: \mathfrak{Z}_{\pi_\gamma}(\mathcal{X}) = \mathbb{C}1$.

2) The roles of *sectors as Micro-Macro boundary* can be seen in *Micro-Macro duality* as a mathematical formulation of "*Quantum-Classical correspondence*" between microscopic *intra-sectorial* & macroscopic *inter-sectorial* levels described by geometrical structures on central spectrum $Sp(3) := Spec(\mathfrak{Z}_\pi(\mathcal{X}))$:

Micro-Macro Duality of Intra- vs. Inter-sectorial levels

$\longleftarrow$ Visible *Macro*		of	$\boldsymbol{Spec} =$	*classifying space*		$\longrightarrow$	*Inter-sectorial*
$\cdots$	γ_N	$\cdots$	*sectors* γ	γ_2	γ_1		$Sp(3)$
	$\vdots$	$\vdots$	$\vdots$	$\vdots$	$\vdots$		*Intra-sectorial*
$\cdots$	π_{γ_N}	$\cdots$	π_γ	π_{γ_2}	π_{γ_1}		$\parallel$
	$\vdots$	$\vdots$	$\vdots$	$\vdots$	$\vdots$		invisible *Micro*

Different sectors: mutually *disjoint* with respect to *unbroken* symmetry, and *connected* by the actions of *broken* symmetries

As explained later, this contrast is shared even by D(H)R theory of *unbroken* symmetry!

4. Emergence of Macro Spec & Symmetry Breaking

3a) *Emergence Process* [Macro $\Longleftarrow$ Micro] of Spec = sector-classifying space via *forcing* along (generic) filters

Mathematically this is controlled by *Tomita theorem* of integral decomposition of a Hilbert bimodule $_{\pi(\mathcal{X})''}\widetilde{\mathcal{X}}_{L^\infty(E_{\mathcal{X}})} := \pi(\mathcal{X})'' \otimes L^\infty(E_{\mathcal{X}})$ with left $\pi(\mathcal{X})''$ & right $L^\infty(E_{\mathcal{X}}, \mu)$ actions, via *central measure* μ supported by *Spec*= supp(μ) = $Sp(3) \subset F_{\mathcal{X}}$: factor states in state space $E_{\mathcal{X}}$ of $\mathcal{X}$.

$\Longrightarrow$ Applications to statistical inference based on large deviation principle [4] and to derivation of Born rule [5].

3b) *Symmetry Breaking & Emergence of Classifying Space*

Sector-classifying space emerges typically from spontaneous breakdown of symmetry of a dynamical system $\mathcal{X} \curvearrowright G$ with action of a group G ("spontaneous" = no changes in dynamics of the system).

4.1. *Symmetry breaking & classifying space*

Criterion for Symmetry Breaking (SB criterion, for short) [1,2]: judged by non-triviality of ***central*** dynamical system $\mathfrak{Z}_\pi(\mathcal{X}) \curvearrowleft G$ associated with the original one $\mathcal{X} \curvearrowleft G$.

I.e., symmetry G is ***broken in sectors*** $\in Sp(\mathfrak{Z})$ ***with non-trivial responses to central G-action***.

G-transitivity assumption with ***unbroken*** subgroup H in broken G leads to sector-classifying space in a specific form of homogeneous space G/H.

$\implies$ ***Classical geometric*** structure on G/H arises physically from ***emergence*** process via ***condensation*** of a family of ***degenerate vacua***, each of which is mutually distinguished by condensed values $\in Sp(\mathfrak{Z}) = G/H$.

In this way, ∞-number of low-energy quanta are condensed into geometry of classical Macro objects $\in G/H$.

4.2. *Sector bundle & logical extension from constants to variables*

In combination with sector structure $\hat{H}$ of unbroken symmetry H (à la DHR-DR theory), total sector structure due to this symmetry breaking is described by a ***sector bundle*** $G \underset{H}{\times} \hat{H}$ with fiber $\hat{H}$ over base space G/H consisting of "***degenerate vacua***" [1,2].

When this geometric structure is established, all the physical quantities are ***parametrized by condensed values of order parameters*** $\in G/H$
$\implies$ "***Logical extension***" [6] of ***constants*** (= global objects) into ***sector-dependent function objects*** (: origin of ***local gauge*** structures)

5. *G/H as Symmetric Space*

This homogeneous space G/H is shown to be a ***symmetric space*** with Cartan involution as follows [IO, in preparation].

Lie-bracket relations $[\mathfrak{h}, \mathfrak{h}] \subset \mathfrak{h}$, $[\mathfrak{h}, \mathfrak{m}] \subset \mathfrak{m}$ hold for Lie structures $\mathfrak{g}, \mathfrak{h}, \mathfrak{m}$ of $G, H, M := G/H$. If $[\mathfrak{m}, \mathfrak{m}] \subset \mathfrak{h}$ is verified, M becomes a symmetric space (at least, locally) equipped with Cartan involution $\mathcal{I}$ with eigenvalues $\mathcal{I} \restriction_\mathfrak{h} = +1$ & $\mathcal{I} \restriction_\mathfrak{m} = -1$:

This property $[\mathfrak{m}, \mathfrak{m}] \subset \mathfrak{h}$ follows from the relation: $[\mathfrak{m}, \mathfrak{m}] = $ ***holonomy*** associated with an infinitesimal loop in ***inter-sectorial space*** $M = Sp(3)$ along ***broken direction***. Since $[\mathfrak{m}, \mathfrak{m}]$= effect of ***broken*** G transformation along an infinitesimal loop $\overset{\gamma}{\circlearrowleft}$ on M starting from and returning to the same point $\gamma \in M$. Thus, $\mathfrak{m}$-component in $[\mathfrak{m}, \mathfrak{m}]$ is absent by the above SB criterion, and hence, $M = G/H = Sp(3)$ is a symmetric space (at least, locally).

Example 1): Lorentz boosts

Typical example of this sort can be found for Lorentz group $\mathcal{L}_+^\uparrow =: G$, rotation group $SO(3) =: H$, $G/H = M \cong \mathbb{R}^3$: symmetric space of Lorentz frames connected by Lorentz boosts.

For $\mathfrak{h} := \{M_{ij}; i, j = 1, 2, 3, i < j\}$, $\mathfrak{m} := \{M_{0i}; i = 1, 2, 3\}$,
$[\mathfrak{h}, \mathfrak{h}] = \mathfrak{h}$, $[\mathfrak{h}, \mathfrak{m}] = \mathfrak{m}$, $[\mathfrak{m}, \mathfrak{m}] \subset \mathfrak{h}$: verified by the basic Lie algebra structure:
$$[iM_{\mu\nu}, iM_{\rho\sigma}] = -(\eta_{\nu\rho}iM_{\mu\sigma} - \eta_{\nu\sigma}iM_{\mu\rho} - \eta_{\mu\rho}iM_{\nu\sigma} + \eta_{\mu\sigma}iM_{\nu\rho}).$$

In contrast to the usual interpretation of Lorentz invariance, ***unbroken*** Lorentz boosts $\mathfrak{m}$ is ***speciality of the vacuum situation***, due to such results as Borcher-Arveson theorem (: Poincaré generators can be physical observables only in vacuum representation) & spontaneous breakdown of Lorentz boosts at $T \neq 0K$ [7]. In this sense, Lorentz frames $M \cong \mathbb{R}^3$ with [boost, boost] = rotation, give a typical example of symmetric space structure emerging from symmetry breaking.

Example 2): Along this line, ***chiral symmetry*** with current algebra structure $[V, V] = V, [V, A] = A, [A, A] = V$ and ***conformal symmetry*** also provide typical examples.

Example 3): 2nd Law of Thermodynamics

Physically more interesting example can be found in ***thermodynamics***: 1st law of thermodynamics $\Longrightarrow \Delta'Q \hookrightarrow \Delta E = \Delta'Q + \Delta'W \twoheadrightarrow \Delta'W$: exact sequence corresponding to $\mathfrak{h} \hookrightarrow \mathfrak{g} \twoheadrightarrow \mathfrak{m} = \mathfrak{g}/\mathfrak{h}$.

With respect to Cartan involution with $+$ assigned to heat production $\Delta'Q$ and $-$ to macroscopic work $\Delta'W$, the holonomy $[\mathfrak{m}, \mathfrak{m}] \subset \mathfrak{h}$ corresponding to a loop in the space M of thermodynamic variables becomes just

Kelvin's version of 2nd law of thermodynamics

namely, holonomy $[\mathfrak{m}, \mathfrak{m}]$ in the cyclic process with $\Delta E = \Delta'Q + \Delta'W = 0$, describes heat production $\Delta'Q \geq 0$: $-\Delta'W = -[\mathfrak{m}, \mathfrak{m}] = \Delta'Q > 0$ (from system to outside)

6. Sector Bundle & Holonomy along Goldstone Condensates

In use of sector bundle $\hat{H} \hookrightarrow G \underset{H}{\times} \hat{H} \twoheadrightarrow G/H$, physical origin of space-time concept can be seen in its ***physical emergence process*** [8].

For simplicity, we assume here that a group G of broken internal symmetry be extended by a group $\mathcal{R}$ of space-time symmetry (typically translations) into a larger group $\Gamma = \mathcal{R} \dot{\times} G$ defined by a semi-direct product of $\mathcal{R}$ & G with $\Gamma/G = \mathcal{R}$.

In this case, the sector bundles have a double fibration structure:

$$\hat{H} \hookrightarrow G \underset{H}{\times} \hat{H} \hookrightarrow \Gamma \underset{G}{\times} (G \underset{H}{\times} \hat{H}) = \Gamma \underset{H}{\times} \hat{H}$$
$$\downarrow \qquad\qquad\qquad \downarrow$$
$$G/H \qquad\qquad \Gamma/G = \mathcal{R}$$

Thus we have three different axes on different levels in Spec:
a) sectors $\hat{H}$ of *unbroken* symmetry H,
b) deg. vacua $G/H = M$ due to *broken internal* symmetry [1,2],
c) $\Gamma/G = \mathcal{R}$ as emergent *space-time* [8] in broken external symmetry.

These axes arise in a series of structure-group contractions $H \leftarrow G \leftarrow \Gamma$ of principal bundles $P_H \hookrightarrow P_G \hookrightarrow P_\Gamma$ over $\mathcal{R}$, specified by ***solderings*** as bundle sections, $\mathcal{R} \overset{\rho}{\hookrightarrow} P_G/H = P_H \underset{H}{\times} (G/H)$, $\mathcal{R} \overset{\tau}{\hookrightarrow} P_\Gamma/G = P_G \underset{G}{\times} (\Gamma/G) = P_G \underset{G}{\times} \mathcal{R}$, corresponding physically to ***Goldstone modes***:

$$
\begin{array}{ccccc}
P_H & \hookrightarrow & P_G & \hookrightarrow & P_\Gamma \\
H \downarrow & \circlearrowleft & \downarrow H & \circlearrowleft & \downarrow H \\
\mathcal{R} & \overset{\rho}{\hookrightarrow} & P_G/H & \overset{\sigma}{\hookrightarrow} & P_\Gamma/H \\
& \backslash\backslash\circlearrowleft & \downarrow G/H & \circlearrowleft & \downarrow G/H\, . \\
& & \mathcal{R} & \overset{\tau}{\hookrightarrow} & P_\Gamma/G \\
& & & \backslash\backslash\circlearrowleft & \downarrow \mathcal{R} \\
& & & & \mathcal{R}
\end{array}
$$

7. Augmented Algebra as Algebraic Dual of Helgason Duality

From the algebraic viewpoint (dual to **Helgason duality** $K\backslash G \leftrightarrow G/H$:

$\nearrow \quad K\backslash G/H \quad \nwarrow$

$K\backslash G \quad \leftrightarrow \quad G/H$, with Radon transforms & **Hecke algebra** $K\backslash G/H$),

$\nwarrow \quad G \quad \nearrow$

the essence of the relevant structures can be viewed as the "*stereographic*" **extension** of such *planar* diagrams as controlling "augmented algebras" [1] of crossed products to describe symmetry breaking:

$$
\begin{array}{ccc}
{}_{G/H}\swarrow \mathcal{X}^H = \tilde{\mathcal{X}}^G \searrow_H & {}_{\mathcal{R}}\swarrow \mathcal{O}_\rho = \mathcal{O}_d^H \searrow_H & \text{[same sort} \\
\tilde{\mathcal{X}}^H \quad \Downarrow \quad \mathcal{X} & \mathcal{A}(\mathcal{R}) \quad \Downarrow \quad \mathcal{O}_d & \text{of lines are} \\
{\downarrow}_H \searrow\searrow \tilde{\mathcal{X}} \swarrow_{G/H} \downarrow \quad \rightleftarrows \quad {\downarrow}_H \searrow\searrow \mathcal{X}(\mathcal{R}) \swarrow_{\mathcal{R}} \downarrow & : \text{in the same} \\
\downarrow \swarrow \quad \Downarrow \searrow\searrow \downarrow \quad \downarrow \swarrow \quad \Downarrow \searrow\searrow \downarrow & \text{exact seq]} \\
\widehat{H\backslash G} \hookrightarrow \widehat{G} \rightarrow \widehat{H} \quad \widehat{\mathcal{R}} \hookrightarrow \widehat{\Gamma} \twoheadrightarrow \widehat{H} &
\end{array}
$$

Note that push-out diagram shows up here (right) in DR reconstruction [9] of field algebra $\mathcal{X}(\mathcal{R})$ with its internal symmetry unbroken.

8. Symmetric Space Structure & Maxwell Equation

Symmetric space structures of $G/H = M$ & $\Gamma/G = \mathcal{R}$ due to symmetry breaking is characterized by the equation of type $[\mathfrak{m}, \mathfrak{m}] \subset \mathfrak{h}$, which connects holonomy $[\mathfrak{m}, \mathfrak{m}]$ (in terms of curvature) with generators $\mathfrak{h}$ of unbroken subgroup.

Note that this feature is shared in common by Maxwell & Einstein equations of electromagnetism and of gravity, respectively:

LHS: (curvature $F_{\mu\nu}$ or $R_{\mu\nu}$) = (source current J_μ or $T_{\mu\nu}$) : RHS.

According to the second Noether theorem (developed in the theory of invariants), Maxwell equation is an identity following from the invariance of action integral under space-time dependent transformations. In contrast, however, *no such classical quantities as action integrals nor Lagrangian densities* are available in our algebraic & categorical formulation of quantum fields.

8.1. *Spectral functor in Doplicher-Roberts reconstruction of symmetry*

The expected roles of action integral are to determine representation contents of a theory. In Doplicher & Roberts (DR) reconstruction [9], this can be substituted by categorical data concerning Galois group in terms of DR category $\mathcal{T}$ of modules of local excitations:

$\mathrm{Obj}(\mathcal{T})$: local endomorphisms $\rho \in End(\mathcal{A})$ of observable algebra $\mathcal{A}$, selected by DHR localization criterion $\pi_0 \circ \rho \restriction_{\mathcal{A}(\mathcal{O}')} \cong \pi_0 \restriction_{\mathcal{A}(\mathcal{O}')}$,

$\mathrm{Mor}(\mathcal{T})$: $T \in \mathcal{T}(\rho \leftarrow \sigma) \subset \mathcal{A}$ intertwining $\rho, \sigma \in \mathcal{T}$: $\rho(A)T = T\sigma(A)$.

In this context, the group H of unbroken internal symmetry is identified with the group $H = End_\otimes(V)$ of unitary tensorial (=monoidal) natural transformations $u : V \leftarrow V$ with the spectral functor $V : \mathcal{T} \hookrightarrow Hilb$ to embed $\mathcal{T}$ into category $Hilb$ of Hilbert spaces with morphisms as bounded linear maps.

8.2. *Spectral functor in category & its local gauge invariance*

Noting the commutativity diagram,

$$v_\rho W(T) = V(T)v_\sigma : \begin{array}{ccc} V(\rho) & \overset{v_\rho}{\leftarrow} & W(\rho) \\ V(T) \uparrow & \circlearrowleft & \uparrow W(T) \\ V(\sigma) & \underset{v_\sigma}{\leftarrow} & W(\sigma) \end{array},$$

to define a natural transformation $v : V \leftarrow W$ from a functor W to another V with $T \in \mathcal{T}(\rho \leftarrow \sigma)$ [10], we re-interpret it as a categorical definition of a *local gauge transformation* $W \overset{\tau_v}{\to} \tau_v(W) = V$ of a functor W to V on the basis of definition:

$$\tau_v(W)(T) := v_\rho W(T) v_\sigma^{-1} \quad \text{for } T \in \mathcal{T}(\rho \leftarrow \sigma).$$

Note that similar formulae appear for gauge links in lattice gauge theory.

Then, the commutativity, $u_\rho V(T) = V(T)u_\sigma$ for $u \in End_\otimes(V)$, can be interpreted as *local gauge invariance* $\tau_u(V) = V$ of the functor V under *local gauge transformation* $V \to \tau_u(V)$ induced by a natural transformation $u \in H = End_\otimes(V)$.

8.3. *Local gauge invariance & Maxwell equation*

In the original DR theory, local endomorphisms $\rho \in \mathcal{T} \subset End(\mathcal{A})$ have, un-fortunately, been regarded as *global* constant objects, owing to the emphasis on space-time transportability[a], and hence, the left-right difference of u_ρ and u_σ in $\tau_u(V)(T) := u_\rho V(T)u_\sigma^{-1}$ has not been recognized as important signal of local gauge structures.

---

[a]This has led to the mathematical definition of "sectors" of $\mathcal{A}$ by $End(\mathcal{A})/Inn(\mathcal{A})$.

From the viewpoint of forcing method, however, the essential features of logical extension *from constants to variables* [6] naturally lead to the interpretation of $\tau_u(V)(T) = u_\rho V(T)u_\sigma^{-1} = V(T)$ as the characterization of local gauge invariance of the functor V under local gauge transformation $u : \mathcal{T} \ni \rho \longmapsto u_\rho$. This is in harmony also with the alternative formulation of principal bundles in terms of group-valued Čech cohomologies.

9. Second Noether Theorem & Maxwell Equation

To adapt the roles of DR category $\mathcal{T} \subset End(\mathcal{A}) = End(\mathcal{X}^H)$ in determining the factor spectrum $Sp(3(\mathcal{X}^H)) = \hat{H}$ to our present purpose, we need to replace $\mathcal{T}$ by $\widetilde{\widetilde{\mathcal{T}}} = End(\widetilde{\widetilde{\mathcal{X}}}^H)$ with $\widetilde{\widetilde{\mathcal{X}}} = \mathcal{X}^H \rtimes \hat{\mathcal{R}}$ and with $\Gamma/G = \mathcal{R}$(: spacetime) in the two-step construction of augmented algebras associated with the series of group extensions: unbroken $H \hookrightarrow$ broken internal $G \hookrightarrow$ broken external Γ.

By repeating the categorical formulation of $End_\otimes(V : \mathcal{T} \hookrightarrow Hilb)$ with $\mathcal{T}$ and V replaced by $\widetilde{\widetilde{\mathcal{T}}}$ and $\widetilde{\widetilde{V}}$, respectively, we can reproduce the essence of the second Noether theorem to connect the local gauge invariance and Maxwell equation. In this context, the second Noether theorem can be generalized into a form with three type arguments, $x \in \mathcal{R}, \xi \in G/H, a \in \hat{H}$.

For simplicity, we reproduce its standard form with infinitesimal local gauge transformation $\delta_\Lambda \varphi^a(x) = G^a(x) \cdot \Lambda(x) + T^{a\mu}(x) \cdot \partial_\mu \Lambda(x)$ of fields $\varphi^a(x)$ specified by an "infinitesimal parameter" $\Lambda = \Lambda(x)$ of a natural transformation depending on sector parameter $x \in \mathcal{R}$. Then Maxwell-type equation holds identically,

$$\partial_\nu K^{\nu\mu} + J^\mu = 0,$$

with $K^{\nu\mu}$ and J^μ defined in relation with the "infinitesimal transforms" of spectral functor V:

$$K^{\nu\mu} := T^{a\mu} \frac{\partial}{\partial(\partial_\nu \varphi^a)} V,$$

$$J^\mu := T^{a\mu}\left(\frac{\partial}{\partial \varphi^a} - \frac{\partial}{\partial(\partial_\nu \varphi^a)}\right) V + G^a \frac{\partial}{\partial(\partial_\mu \varphi^a)} V.$$

Choosing $\xi \in G/H$ as the parameter-dependence of local gauge transformations, we can incorporate the low-energy theorem (with "soft pions") due to symmetry breaking in the present context.

In the case with $a \in \hat{H}$, we note that the recovered group H of unbroken symmetry is compact in DR theory [9] which implies that the group dual $\hat{H}$

of sector parameters is discrete. While it seems difficult to adapt this case to the standard formulation of the second Noether theorem in terms of differential operations, we expect some interesting lessons to be learned from the attempt to unify it in the present context.

References

1. Ojima, I., A unified scheme for generalized sectors based on selection criteria –Order parameters of symmetries and of thermality and physical meanings of adjunctions–, Open Systems and Information Dynamics, **10**, 235-279 (2003) (math-ph/0303009)
2. Ojima, I., Temperature as order parameter of broken scale invariance, Publ. RIMS (Kyoto Univ.) **40**, 731-756 (2004) (math-ph0311025).
3. Ojima, I., Micro-macro duality in quantum physics, 143-161, Proc. Intern. Conf. "Stochastic Analysis: Classical and Quantum", World Sci., 2005, arXiv:math-ph/0502038; Lévy Process and Innovation Theory in the context of Micro-Macro Duality, 15 December 2006 at The 5th Lévy Seminar in Nagoya, Japan.
4. Ojima, I. and Okamura, K., Large deviation strategy for inverse problem I & II, Open Sys. Inf. Dyn., **19**, 1250021 & 1250022 (2012).
5. Ojima, I., Okamura, K. and Saigo, H., Derivation of Born rule from algebraic and statistical axioms. **21** No. 3 1450005 (2014).
6. Ojima, I. and Ozawa, M., Unitary Representations of the Hyperfinite Heisenberg Group and the Logical Extension Methods in Physics, Open Systems and Information Dynamics **2**, 107-128 (1993).
7. Ojima, I., Lorentz invariance vs. temperature in QFT, Lett. Math. Phys. **11**, 73-80 (1986).
8. Ojima, I., Space(-Time) Emergence as Symmetry Breaking Effect, Quantum Bio-Informatics IV, 279 - 289 (2011) (arXiv:math-ph/1102.0838 (2011)).
9. Doplicher, S. and Roberts, J.E., Why there is a field algebra with a compact gauge group describing the superselection structure in particle physics, Comm. Math. Phys. **131**, 51-107 (1990).
10. Mac Lane, S., *Categories for the working mathematician*, Springer-Verlag, 1971.

Quantum Bio-Informatics VI
© 2020 World Scientific Publishing Co. Pte. Ltd.
pp. 139–148

USING HOMOMORPHIC ENCRYPTION SYSTEM FOR HARASSMENT CONTROL WITH COMPLETE RESPECT OF PRIVACY

MASSIMO REGOLI

*Ricercatore DICII, Università di Roma "Tor Vergata", Via del Politecnico, 2
Roma, Italy
E-mail: regoli@uniroma2.it*

In this article we will discuss how it is possible to apply homomorphic encryption, and in particular the Pailler Cryptosystem, to problems regarding the tracking of individuals subjected to some restrictions on freedom of movement, while maintaining respect for their privacy. In particular, we discuss the problem of the minimum distance to be maintained in cases of harassment between a Victim V and Stalker S and how to create a control mechanism without the requirement of any of the parties to have the complete knowledge of the positions of V and S. Only in case of need of intervention, these sensitive information, through a collaboration between some other parties, can be made accessible.

1. Introduction

1.1. *Some statistics*

Nowadays between 15 and 76 percent of women are targeted for physical and/or sexual violence in their lifetime, according to the available country data and most of this violence takes place within intimate relationships, with many women (ranging from 9 to 70 percent) reporting their husbands or partners as the perpetrator. Across the 28 States of the European Union, a little over one in five women has experienced physical and/or sexual violence from a partner (European Union Agency for Fundamental Rights, 2014). [a]

A **restraining order** or **order of protection** is a form of court order that requires a party to do, or to refrain from doing, certain acts. As an example the offender shall be sentenced to not approach the Victim to less than 500 meters. But how to check this?

---

[a]See for further information: http://www.endvawnow.org/en/articles/299-fast-facts-statistics-on-violence-against-women-and-girls-.html

In principle it would be possible, with the agreement of the Victim and Stalker, to control the relative distance, using a continuous communication of their position and the computation of the relative distance based on a GPS receiver. If this distance decreases and falls below the threshold, authorities can react accordingly.

1.2. *Privacy issues*

Laws on privacy are an issue that could, however, prevent a constant monitoring of the positions of the actors involved. In fact suppose to implement a scheme like the following:

- Be V the Victim, S the Stalker and P the Policeman, and:

$$C_V = (\lambda_V, \phi_V), C_S = (\lambda_S, \phi_S)$$

  respectively the position of the Victim and the Stalker
- then periodically V and S send their positions C_V and C_S to P
- this allows P to compute the distance between them using an appropriate metric:

$$D_n(V, S) = D_n(C_V, C_S)$$

- If the distance is less than the threshold Policeman can react

In this schema, protection of privacy is the issues: in fact the Policeman P knows the positions of Stalker S and Victim V, moreover the communication can be intercepted and many countries have laws that protect sensitive information, such as the location of an individual, as a private property. At last an individual who suffers an offense does not necessarily want to be tracked as it moves even by a Policeman and, in some cases, this is true also for the Stalker.

Apply cryptographic methods could solve some of the problems, but not all of them. In fact Policeman P to compute the distance between S and V should have knowledge of their absolute position: privacy is not fully respected.

2. An approach to a solution

2.1. *Homomorphic cncryption*

The introduction of a fourth actor (an *Arbiter*) and the use of the *homomorphic encryption* could solve all the problems about privacy.

In fact, suppose we have an encryption algorithm able to compute distances between encrypted point on a plane (or on a sphere) in an homomorphic way, then our idea is to follow the next schema (see also figure 1):

(1) Be V the Victim, S the Stalker P the Policeman, A an Arbiter, and be:

$$C_V = (\lambda_V, \phi_V), C_S = (\lambda_S, \phi_S)$$

respectively the Coordinates of Victim and Stalker

(2) V and S encrypt their data position using a common pre-shared public key k using a fully or a partial **homomorphic encryption algorithm**

(3) then periodically V and S send their encrypted coordinates $\tilde{C}_V = \left(\tilde{\lambda}_V, \tilde{\phi}_V\right)$ and $\tilde{C}_S = \left(\tilde{\lambda}_S, \tilde{\phi}_S\right)$ to P

(4) P compute the encoded distance $\tilde{d}_{VS}$ between V and S

$$\tilde{d} = \tilde{d}_{VS} = d(\tilde{C}_V, \tilde{C}_S)$$

using the public key k

(5) P sends $\tilde{d}$ to A

- **nota bene**: P doesn't know the secret key associated with the public key k so he doesn't know the absolute position of V and S, he can only compute mathematical operation on encrypted value using k: in this case privacy is completely respected

(6) A receives $\tilde{d}$, decrypt it using the secret key associated to the public key k and eventually finalize the computation.

- **nota bene**: A knows only the relative distance of V and S but doesn't know their absolute position: also in this case privacy is completely respected

(7) If the distance is less than the threshold he warns the Policeman P for a first reactions (i.e. the Cop could send a message to Stalker asking him to move away from actual position)

(8) In extreme cases A may ask to P the absolute encrypted position of the Stalker to decipher it and instruct any request for law enforcement on the site (in many countries, for immediate emergency, the privacy of citizens may be violated by authority).

142

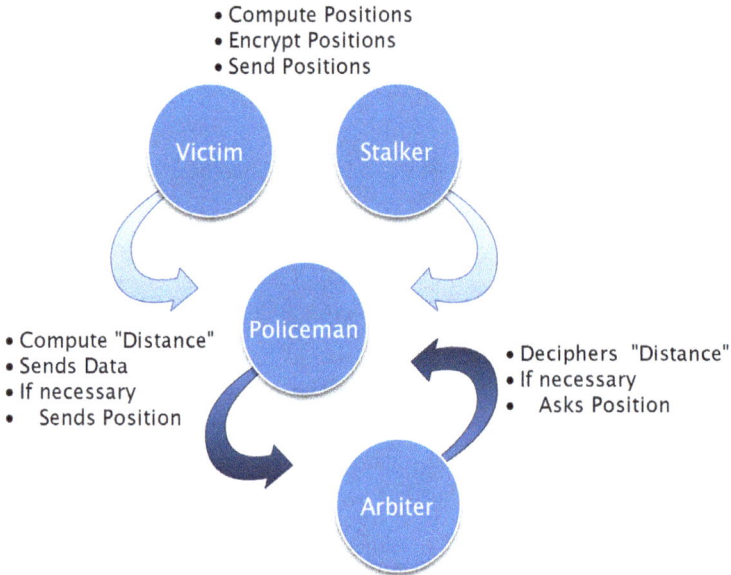

Figure 1.   The exhaustive communication diagram between the four actors.

2.2. *Tools*

So homomorphic encryption could be valid solution to solve this problem but actual questions are:

- What is homomorphic encryption?
- How many different types exist?
- Which one is useful for our scenario?

2.2.1. *Definitions*

Informally speaking, a homomorphic cryptosystem is a cryptosystem with the additional property that there exists an efficient algorithm to compute an encryption of the sum and/or the product, of two messages given the public key and the encryptions of the messages but not the messages themselves. Then we can have:

- additively homomorphic,
- multiplicatively homomorphic,
- full homomorphic

cryptosystems.

Just to fix the idea:

$$E(a) + E(b) = E(a + b) \tag{1}$$

$$E(a)E(b) = E(ab) \tag{2}$$

2.2.2. *Partially vs Fully Homomorphic Encryption*

Partially Homomorphic Encryption (PHE) has the opportunity to compute one type of operation over encrypted values. For instance, if you know $E(m_1)$ and $E(m_2)$ you can, without knowing the private key, compute $E(m_1 * m_2)$.

Several efficient algorithms which allow for that are known, especially ElGamal or RSA (which allows you to multiply encrypted messages) and the Paillier cryptosystem (which allows you to add encrypted messages). Partially homomorphic encryption is useful for some protocols, e.g. electronic voting (the idea is that you can *sum* together encrypted votes, and decrypt the result at the end).

On the other hand Fully Homomorphic Encryption (FHE) can computes two types of operations over encrypted values: from $E(m_1)$ and $E(m_2)$, you can compute $E(m_1 * m_2)$ and $E(m_1 + m_2)$.

Unfortunately, the best known fully homomorphic encryption algorithms (derived from Gentry's work in 2009) are awfully slow and inefficient, making it not worth the effort (i.e. the whole Amazon S3 cloud could not compute homomorphically faster than what a single smartcard could do without the encryption).

2.2.3. *Pailler Cryptosystem*

The Paillier Cryptosystem (PC), named after and invented by Pascal Paillier in 1999[3], is a probabilistic asymmetric algorithm for public key cryptography. The problem of computing *n-th* residue classes is believed to be computationally difficult. Moreover the scheme is an additive homomorphic cryptosystem; this means that, given only the public-key and the encryption of m_1 and m_2, one can compute the encryption of $m_1 + m_2$ and, of course, $m_1 - m_2$.

The PC system is the follow:

- p, q two large prime numbers such that $gcd(pq, (p-1)(q-1)) = 1$
- $n = pq$

- $\lambda = lcm(p-1, q-1)$
- $g \in \mathbb{Z}_{n^2}^*$ such that n divides order of g
- Public Key is: (n, g)
- Private Key is: (λ, μ) where $\mu = \left(\dfrac{(g^\lambda \mod n^2) - 1}{n} \right)^{-1} \mod n$
- **Encryption**:
  - $m \in \mathbb{Z}_n$ a message
  - $r \in \mathbb{Z}_n^*$ a random number
  - the ciphertext will be:

$$c = g^m r^n \mod n^2$$

- **Decryption**:
  - $c \in \mathbb{Z}_{n^2}^*$ a ciphertext
  - the plaintext will be:

$$m = \left(\frac{(c^\lambda \mod n^2) - 1}{n} \right) \mu \mod n$$

2.2.4. *thep library*

thep project aims to provide homomorphic encryption libraries to developers so they can in turn create privacy and confidentiality aware software. Currently the code implements the Paillier cryptosystem in Java, along with it's homomorphic operations and key generation [b].

The source code below shows an example of using the library:

```
public static void main(String[] args) {
    // Create a random private key of 128 bits
    PrivateKey privkey = new PrivateKey(128);

    // Fetch the corresponding public key
    PublicKey pubkey = privkey.getPublicKey();

    EncryptedInteger enc_a =
      new EncryptedInteger(new BigInteger("155"), pubkey);
    System.out.println("Enc_pub(a) = " + enc_a.getCipherVal());

    EncryptedInteger b = enc_a.add(enc_a);
```

---

[b] https://code.google.com/p/thep

```
System.out.println
   ("b = Enc_pub(a) + Enc_pub(a) = " + b.getCipherVal());

System.out.println
   ("Dec_priv(b) = " + b.decrypt(privkey));
}
```

2.3. Formulas

For some reasons we decide to adopt a PHE schema. Specifically the pailler schema implemented in the thep library.

But to calculate the distance between two points on the earth's sphere we need both sums and multiplications (*and some other mathematical gadgets*). As an example, the Haversine formula can give the great-circle distances between two points on a sphere from their longitudes and latitudes:

$$d = 2r \arcsin\left(\sqrt{\sin^2\left(\frac{\phi_S - \phi_V}{2}\right) + \cos(\phi_S)\cos(\phi_V)\sin^2\left(\frac{\lambda_S - \lambda_V}{2}\right)}\right) \tag{3}$$

$$= 2r \arcsin\left(\sqrt{\sin^2\left(\frac{\Delta\phi}{2}\right) + \cos(\phi_S)\cos(\phi_V)\sin^2\left(\frac{\Delta\lambda}{2}\right)}\right) \tag{4}$$

2.4. Approximations

The following scheme:

- V and S sends their relative position chyphered $\tilde{\phi}$, $\tilde{\lambda}$.
- Cop computes the encrypted difference $\Delta\tilde{\phi}$, $\Delta\tilde{\lambda}$ and sends it to Arbiter
- Arbiter A decode $\Delta\phi$, $\Delta\lambda$ and can compute.

$$\sin^2\left(\frac{\Delta\phi}{2}\right) \quad \text{and} \quad \sin^2\left(\frac{\Delta\lambda}{2}\right)$$

there is a privacy problem because A can't compute:

$$\gamma = \cos(\phi_1)\cos(\phi_2) \tag{5}$$

because if the Arbiter knows $\cos(\phi_*)$ he might know the latitude of V and S and this would undermine their privacy.

2.4.1. *Mercator projection*

Haversine formulas are not compatible with a PHE algorithm. But, thanks to the Mercator projection, we can consider a sphere like a plane:

$$\begin{cases} x = R\left(\lambda - \lambda_0\right), \\ y = R\ln\left(\tan\left(\frac{\pi}{4} + \frac{\phi}{2}\right)\right) \end{cases} \tag{6}$$

where in the first equation of (6), λ_0 is the longitude of an arbitrary central meridian usually, but not always, that of Greenwich (i.e., zero), R is the earth radius.

And the inverse transformations are:

$$\begin{cases} \lambda = \lambda_0 + \frac{x}{R} \\ \phi = 2\tan^{-1}\left(e^{\frac{y}{R}}\right) - \frac{\pi}{2} \end{cases} \tag{7}$$

This consideration allows us to use a metric on the plane. For example we can consider the euclidean distance (eq. (8)) or the manhattan (eq. (9)) distance.

$$d_E(S, V) = \sqrt{(x_S - x_V)^2 + (y_S - y_V)^2} \tag{8}$$

$$d_M(S, V) = |x_S - x_V| + |y_S - y_V| \tag{9}$$

2.4.2. *Definitive formulas*

Let

$$E(x, k) : \mathbb{Z}_| \to \mathbb{Z}_|$$

and

$$D(x, k) : \mathbb{Z}_| \to \mathbb{Z}_|$$

respectively an homomorphic encryption function and the respective decryption function with secret key k, then, in this scenario, V and S compute formulas expressed in equation (6) then, using the secret key S_k, they encrypt their data:

$$\begin{aligned} \tilde{x} &= E(x, k) = E\left(R\left(\lambda - \lambda_0\right), k\right) \\ \tilde{y} &= E(y, k) = E\left(R\ln\left(\tan\left(\frac{\pi}{4} + \frac{\phi}{2}\right)\right), k\right) \end{aligned} \tag{10}$$

then V and S send results to P. At this point P can calculate:

$$\begin{aligned} \Delta\tilde{x} &= (\tilde{x}_V - \tilde{x}_S) \\ \Delta\tilde{y} &= (\tilde{y}_V - \tilde{y}_S) \end{aligned} \tag{11}$$

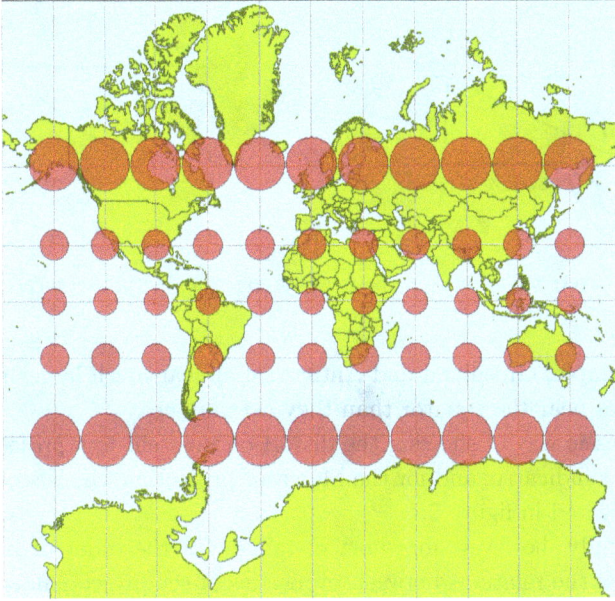

Figure 2. Tissot's indicatrix for Mercator projection

using the relative homomorphic operator and the public key P_k. Then he sends the results $\Delta\tilde{x}, \Delta\tilde{y}$ to A.

Last step A decrypt data using the secret key S_k:

$$\Delta x = D(\Delta\tilde{x}, k)$$
$$\Delta y = D(\Delta\tilde{y}, k)$$

(12)

and then he can calculates:

$$d(S, V) - \sqrt{\Delta x^2 + \Delta y^2}$$

(13)

using Euclidean metric or another equivalent metric. Finally A reacts if $d(S, V) < t$ where t is a *threshold* set beforehand. Here reacting means ask to P the *encrypted values* calculated by V and S in equation (11).

2.5. *Notes on Mercator projection*

Like all projections of the terrestrial sphere on a plane the formulas Mercator present some distortions.

In fact the Mercator projection distorts the size of objects as the latitude increases from the Equator to the Poles, where the scale becomes infinite.

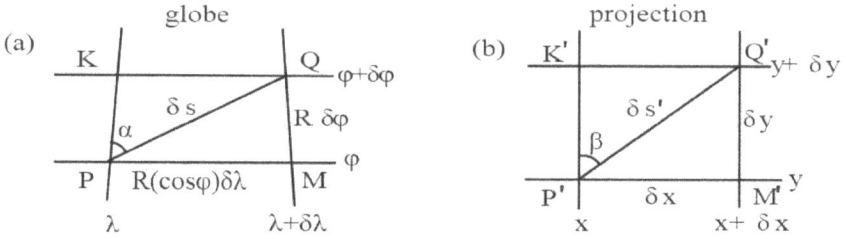

Figure 3. Corresponding measures between the sphere and its projection according with Mercator

So, for example, Greenland and Antarctica appear much larger relative to land masses near the equator than they actually are.

The classic way of showing the distortion inherent in a projection is to use Tissot's indicatrix and for the Mercator projection, the Tissot's indicatrix is indicated in figure 2.

Thankfully, however, for short distances (of the order of hundred of kilometers), the relations between properties of the projection, such as the transformation of angles and the variation in scale, follow from the geometry of corresponding small elements on the globe and map, so in case of short distances the differences between δs and $\delta s'$ if figure are small (see fig. 3).

References

1. *Titolo della tesi*, D. Poto, 2014, Graduation Thesis
2. *A Fully homomorphic encryption scheme*, C. Gentry, 2009, DEGREE OF DOCTOR OF PHILOSOPHY
3. *Public-Key Cryptosystems Based on Composite Degree Residuosity Classes*. P. Paillier, 1999, EUROCRYPT. Springer. pp. 223-238. doi:10.1007/3–540–48910–X_16

Quantum Bio-Informatics VI
© 2020 World Scientific Publishing Co. Pte. Ltd.
pp. 149–161

STOCHASTIC SYSTEMS DEPENDING ON THE NEW NOISE AND THE MULTIPLICITY

SI SI

Emeritus Professor, Aichi Prefectural University, 1522-3, Ibaragabasama,
Nagakute, Aichi 480-1198, Japan
E-mail: ohnbin@gmail.com

Dedicated to Professor Masanori Ohya

Gaussian white noise $\dot{B}(t)$ and Poisson noise $\dot{P}(t)$ are well known typical noises. Besides these two noises, there is one more noise $P'(\lambda)$, a new noise depending on space variable see, for example, [10] and [11].

By using the concept of multiplicity and taking the translations of the parameter of the generating function of lattice type distributions associated with additive process we obtain a compound Poisson process which remains within the unit multiplicity. And then we know that ψ-function of a compound Poisson process admits invariance under the Affine transform. Consequently, we give the general form of $\psi(z)$ from which we can revisit the decomposition of Lévy process.

1. Introduction

Following the **reductionism** for stochastic systems, we know basic systems of idealized elemental random variables. We are interested in the case where the system involves continuously many members. This implies that each member can not be an ordinary random variables, but be really idealized (generalized) elemental random variables (i.e.r.v.)s.

Such a system can be parametrized by a variable running through R^1 or its interval. We have shown that the parameter can be considered as time or space.

With such choice of parameter, we may request that the probability distribution of a system is invariant under the shift of the parameter space, namely "stationary". The term of "elemental" actually means that each member is atomic in probability distribution.

We have so far reminded the results that have been established. A stochastic system satisfying the above conditions is simply called a **noise**.

See, for example, the note presented at the Singapore workshop on IDAQP held in March 2014.

In this paper we shall show that there are three kinds of noises which are extremal. We show a list of them:

i) depending on time t : there are two extremal processes which are Gaussian $\dot{B}(t)$, Poisson type $\dot{P}(t)$, where $B(t)$ is a Brownian motion and $P(t)$ is a Poisson process with unit intensity. The dot means the time derivative.
ii) depending on the space variable λ : that is Poisson $P'(1, \lambda)$, simply written as $P'(\lambda)$, where $P(1, \lambda)$ is a Poisson process with intensity λ (which can be regarded as a space variable) evaluated at $t = 1$. The P' means $\frac{d}{d\lambda} P$.

2. Construction of noises, revisited

We shall recall the construction of noises and add some new viewpoint.

Take the parameter space to be the time running through R^1. Remind the stationarity to take the unit interval $[0, 1]$ on which a system of idealized elemental random variables is defined. As the actual method of construction, we take successive approximation.

Our idea will be explained quickly as follows.

Case I.

1) Divide the unit interval into 2^n subintervals and associate independent identically distributed random variables $X_k^n, 1 \leq k \leq 2^n$, which is denoted by $\mathbf{X^n}$. Each X_k^n should have the third order moment and unit variance. This is the n-th step, where we have the probability distribution μ_n.
2) The $(n + 1)$-th step is carried on in a same manner only with the restriction that $X_{n+1}^{2k} + X_{n+1}^{2k+1}$ is the same type as X_n^k in distribution. This is possible if we take an infinitely divisible distribution.
3) It is easy to see that we have a consistent family of probability distributions $\{\mu_n, n \geq 1\}$. Hence, the inductive limit defines a probability measure μ on $(R^\infty, \mathbf{B})$.
4) The central limit theorem guarantees that there exists a Gaussian random variable and further a Brownian motion, by applying the method to various sub-intervals of $[0, 1]$.

5) Although there are much freedom to choose the probability distribution involved in the above steps, we always have a Brownian motion. This can be considered as the extremal limit.

6) Such observations illustrate the following : the probability distribution of a Brownian motion as well as that of white noise can be approximated in a quite natural manner. This is true since the conclusion does not depend on the choice of the probability distribution of X_k^n, and since the partial sum divided by the length of the corresponding subintervals approximates white noise.

Thus, we can say that a Brownian motion or a white noise always appears regardless the approximation, so far as we use the trick above. We may therefore say that Gaussian noise is extremal or sitting on top.

Case II.

There is another case to have an extremal distribution.

Although we use the same method to use the inductive limit as in Case I, atomic distributions are taken in the process of approximation. For one thing, the limit distribution would be atomic.

The probability distribution of X_k^n is chosen so as to be most simple in the sense of probability distribution.

1) Thus, at the n-th step the random variable is subject to a distribution which is atomic, namely it takes only two values, say a_n and b_n with probability $1 - p_n$ and p_n, respectively. To make the matters simple, we may assume $a_n = 0, b_n = 1$. To keep the expectation of the sum $\sum X_k^n$ to be constant, say λ we define $p_n = \lambda 2^{-n}$. Then, we see that the sum $\sum X_k^n$ is subject to the binomial distribution.

2) As the limit as $n \to \infty$ the distribution of the sum tends to a Poisson distribution with intensity λ by the law of small probability. Further as in the case I, we have a Poisson process with intensity λ.

3) By the choice of the probability distributions of X_k^n we understand the limit process, that is a Poisson process. It is extremal.

We can see similar results for partial sums and obtain a Poisson process $P(t, \lambda)$ with intensity λ. Also we have a Poisson noise $\dot{P}(t, \lambda)$.

Obviously, this is another extremal case, sitting not on top, but bottom, where the probability distributions of the sequence of approximating variables are most simple.

Theorem 2.1. *By the successive approximation we have two extremal additive processes, which are Brownian motion and Poisson process, and hence noises as a result of inductive limit. Thus, we have (Gaussian) white noise and Poisson noise at the same time.*

Here is a very important remark. In the course of setting the probability distribution in Case II, we have chosen a simplest distribution. That is a measure supported by two points. The delta measure is avoided to escape from the trivial result. However, there is a freedom to choose a positive number λ which can be arbitrary. It is in agreement with the expectation of the Poisson process $P(t)$, which we have just formed. It may be written as $P(t, \lambda)$. Let it be evaluated at $t = 1$ to have $P(\lambda)$. This fact tells us that the λ is a space variable. Note that it may be considered to be a scale variable, but the term "scale" is saved so that it is used later, for another meaning.

With this observation we now come to the next step.

Case III.

Let t be fixed, say $t = 1$ and define $P(\Delta_k^n)$ by the same manner as in Case II, where $\Delta_k^n = [\frac{k}{2^n}, \frac{k+1}{2^n}]$. Their sum is denoted by $P(n, \lambda)$. Note that $P(\Delta_{2k}^n) + P(\Delta_{2k+1}^n)$ has the same distribution as $P(\Delta_k^{n-1})$. This means, we can see again the inductive limit denoted by $P(d\lambda)$ and $P(\lambda)$. There runs the λ through $[0, 1]$, but it is easy to extend to that defined over $(0, \infty)$.

It is noted that the choice of λ in ii) is quite arbitrary so far as it is positive. It is, in fact, the expectation of the $P(1, \lambda) = P(\lambda)$, which is a space variable.

As in the cases $\dot{B}(t)$ and $\dot{P}(t)$, we can define $P'(\lambda) = \frac{d}{d\lambda} P(\lambda)$ as a generalized (in fact, idealized) random variable. We call the $P'(\lambda)$ the *space noise*.

Note that $P(\lambda)$ is an additive (that is with independent increments) system.

3. The classification of probability distributions

This section is provided as a simple background for the decomposition of a Lévy process.

Two probability distributions of random variables X and $aX + b$ with $a > 0$ are said to be of the *same type*. This definition implies an equivalence relation. Hence, we are given a classification of all probability distributions $F(x), x \in R^1$.

Such a simple statement may be rephrased as follows. Probability distributions on R^1 are classified by the type. This classification is done by the invariance under the subgroup G, in fact normal subgroup of the Affine group $Aff(R^1)$. To be precise, G is formed by matrices

$$\begin{pmatrix} a & b \\ 0 & 1 \end{pmatrix}$$

where $a > 0$ an b is arbitrary.

An obvious note is that all the Gaussian distributions except the trivial one are of the same type. While Poisson distributions with different intensities are of different types.

The notion of the type is extended to stochastic processes by their probability distributions defined on a function space, that are measures introduced on the (sample) function space.

An additional note is that we shall use the group G in the next section, as was mentioned before, in connection with the discussion on Lévy process.

4. Compound Poisson process

By the term "Compound Poisson process" we mean a linear combination of independent Poisson processes of different types.

It is easy to form compound Poisson processes, however it is not an easy work to decompose the compound system. Actually, it may be thought that it is easy to discriminate the components by observing the jumps by instantaneous work, however we have a process following the time as it goes by.

Another difficulty is that we have to find the intensity of a Poisson process, since the intensity is "not" a visualized parameter.

These problems will be discussed with a new setup in the section that follows.

5. Noises

It is difficult to introduce a system of continuously many independent random variables, say parametrized by the time t running through an interval of R^1. To fix the idea, we assume that they are non-trivial and identically distributed. The joint probability distribution of this system is the direct product of continuously many distributions, which can not be separable, so that it is impossible to discuss integrals of Lebesgue type. This is a crucial difficulty.

There is, however, an idea that leads us to overcome this difficulty. It is to take the time derivatives of an additive process, let it be denoted by $Z(t), t \in T$, with separability like continuity in some sense. The set T is taken to be an interval of R^1. The $Z(t)$ has independent increments, so that take the time derivatives $\dot{Z}(t), t \in T$ to have a system of independent random variables. As is expected, we are given as many as continuum.

Now we have to pay a price. Namely, $\dot{Z}(t)$ is no more ordinary random variable, but it is an idealized random variable. Nevertheless, we are happy to have continuously many independent variables, and we shall be in search of an acceptable interpretation to $\dot{Z}(t)$.

Having done this, we proceed to define functions (actually, functionals) of the $Z(t)$. Further we should develop the analysis of those functionals, namely the analysis of functionals of the $\dot{Z}(t)$.

It is now the time to concretize the $Z(t)$ as well as $\dot{Z}(t)$. After P. Lévy's idea we use approximation $\{X_k^n, 1 \leq k \leq 2^n\}$ over the unit time interval $I = [0, 1]$. For each n take the division $[k/2^n, (k+1)/2^n)$. With the subintervals, we associate i.i.d. random variables $X_k^n, 1 \leq k \leq 2^n$.

There are three cases. The first two are extremal in the respective ways and are time-dependent. The third one is, as it were, a by-product.

i) Assume that each random variable has finite 3rd order moment with mean 0 and variance 2^{-n}. Then, the sum $S_n = \sum_1^{2^n} X_k^n$ converges in law to a standard Gaussian random variable as $n \to \infty$ by the central limit theorem. Also, the law of any consecutive partial sum tends to a Gaussian distribution with variance proportional to the length of the subinterval of I.

Such observations illustrate that the probability distribution of a Brownian motion as well as that of white noise can be approximated in a quite natural manner. This is true since the conclusion does not depend on the choice of the probability distribution of X_k^n, and since the partial sum divided by the length of the corresponding subintervals approximates white noise.

Thus, we can say that a Brownian motion or a white noise always appears regardless the approximation, so far as we use the trick used above. We may therefore say that Gaussian noise is extremal or sitting on top.

ii) There is another extremal, sitting not on top, but bottom. It is the Poisson noise.

The setting is the same as above. The probability distribution of X_k^n is chosen so as to be most simple in the sense of probability distribution. Namely, each X_k^n takes only two values, say, 1 and 0 with probability p_n and $1 - p_n$, respectively. To let the variance of the sum S_n converge, we may assume $p_n = 2^{-n}\lambda$ with some positive constant λ. Then, by the law of small probability, we obtain a Poisson distribution. We can see the similar results for partial sums and obtain a Poisson process $P(t, \lambda)$ with intensity λ. Also we have a Poisson noise $\dot{P}(t, \lambda)$.

Obviously, this is another extremal case, where the probability distributions of the sequence of approximating variables are most simple.

iii) It is noted that the choice of λ in ii) is quite arbitrary so far as it is positive. It is, in fact, the expectation of the $P(1, \lambda) = P(\lambda)$, which is a space variable.

Now one may ask if there exists a noise depending on the space variable λ. The answer is yes.

Proposition 5.1. *There is a noise depending on the space variable λ.*

The idea of the proof is as follows. We have established the exact form of Poisson noise $\dot{P}(t, \lambda)$ which is viewed as a generalized stochastic process with independent values at every t. Its characteristic functional $C^P(\xi)$ is formally expressed in the form

$$C^P(\xi) = E(e^{i\langle \dot{P}, \xi \rangle}).$$

For a smooth function ξ it may be expressed in the form

$$C^P(\xi) = E(e^{-i\langle P, \xi' \rangle}).$$

There exists a system of random variables which is a generalized stochastic process depending on λ with independent values at every point λ. This can be proved by the Bochner-Minlos theorem. The generalized process can be regarded as $P'(\lambda)$, the λ-derivative of the $P(\lambda)$. This fact comes from the following considerations.

Take a test function $\xi(\lambda), \lambda \in (0, \infty)$ to have a proposed formula for the characteristic functional $C^P(\xi)$ of a generalized stochastic process $\dot{P}(\lambda)$, where ξ is a test function in a suitable nuclear space E :

$$C^P(\xi) = \exp[\lambda \int (e^{i\xi(t)} - 1)dt].$$

Note that the $\Psi(\xi) = \log C^P(\xi)$ is linear in λ. It can, therefore, be extended like

$$\Psi^\lambda(\eta) = \eta(\lambda) \int (e^{i\xi(u)} - 1)dt.$$

Note that the extension is not simply as a function but as a Ψ function. It is the log of a positive definite functional. Therefore, we obtain a characteristic functional of η to have a generalized stochastic process depending on the space variable λ.

By the Ψ functionals above, we can see a certain **duality** between time t and space λ through the Poisson type noise.

In the approximation of Poisson noise, we used X_k^n subject to distribution on 2 points; 0 and 1 with probability $1 - p_n$ and p_n respectively. We said it is minimal and the law of small probability gives us a Poisson distribution.

Now we say such two values random variable (distribution) is extremal or minimal. It is extremal (or minimal) in the sense of factorization. Namely, a probability distribution F is said to be *divisible* if there are two non-trivial distributions F_1 and F_2 such that

$$F = F_1 * F_2,$$

where $*$ means the convolution. In terms of random variables, for X, Y, Z having the probability distribution F, F_1, F_2, respectively, then X has the same distribution as $Y + Z$, where every random variable and every distribution is non-trivial. Obviously two valued distribution is "not divisible", so that extremal, or minimal.

6. Decomposition of a compound Poisson process

So far as decomposition is concerned, we must pay attention not to scale (or span for lattice type distributions) but on intensity which is understood to be a space parameter as we understood in the construction of noise.

Unfortunately the intensity is not a visualized parameter, while we can observe the scale of sample functions. For example, the ordinary Poisson process $P(t, \lambda)$ has unit scale, or we may say it has a lattice type distribution with unit "span". Multiplication by u to have $uP(t, \lambda)$ which makes no change in probabilistic nature.

Just forget the observation of the scale. Then, what we should do is as follows. We can only be helped by theoretical approach as in the following.

Take the standard lattice type distribution, i.e. Poisson distribution. It defines an extremal (sitting on bottom) distribution. Take the logarithm of generating function, denote $m(t), t \in R^1$, which is of the form

$$e^{\lambda t} - 1.$$

We introduce the scale u which can be also called "span" as a lattice type distribution. We modify to be $e^{\lambda t u} - 1$. Again, take a modification to be lattice type but not necessarily probability distribution; just a sequence of positive numbers with finite sum. This modification is not essential to discuss the distribution of positive numbers.

Finally, the $m(t)$ is modified to be

$$\tilde{m}(t) = e^{tu} \lambda.$$

Now define the Affine group A of operators $g(a, b)$ acting on t-space R^1 such that

$$A = \{g = g(a, b) : a, b \in R^1\}.$$

where

$$g(a, b)t = at + b.$$

In terms of the matrix form, we may write in the form

$$\begin{pmatrix} a & b \\ 0 & 1 \end{pmatrix}$$

$$m(g(a,b)t) = \exp[e^{bu}\lambda e^{tau}] \tag{1}$$

By the action of $g(a,b)$, a Poisson distribution or its generating function changes in such a way that the intensity λ changes to $e^{bu}\lambda$ and the scale u goes to au.

Here is a very important note. The generating function tells us that by the action of the dilation a the intensity *never changes*. That is, the scale parameter u cannot manage the intensity. This fact gives us an important suggestion on the decomposition of a Lévy process.

Now consider a representation of the group A on the set of modified generating functions. First, fix the parameter λ of the generating function. Take finitely many dilations, say a_j's. Then, we may form $\sum_j m(g(a_j,0)t)$ which corresponds to a sum of *independent* random variables. Obviously the sum can not be a linear combination, since we cannot touch the intensity. Also the sum should not be infinite sum for convergency.

Definition 6.1. The number of independent variables as above is called the **multiplicity** of the representation for the λ fixed in advance, that is, point spectrum.

In fact, the multiplicity may be called discrete multiplicity corresponding to the point spectrum of the intensity.

It is easy to establish a correspondence between the generating functions just obtained and Lévy process with the same intensity but different jumps as many as the multiplicity.

Continuous intensity measure

Coming back to the affine group, we now restrict our attention to the shift, i.e. the action of the shift b. Then, the group acts the *generating function* of the standard Poisson distribution with intensity λ which is

$$m(z) = \varphi(z) = \exp[\lambda(e^{tu} - 1].$$

We, therefore, have

$$m(g(0,b)t) = \exp[e^{bu}\lambda(e^{zu} - 1)]. \tag{2}$$

We have to remind the note made before. We may remove the constant 1 in the above two expressions. Indeed, the constant 1 plays the role to make a positive sequence to be a probability distribution, that is the sum to be 1. Also, we can take the exponent, that is the ψ-function.

$$\psi(z) = \lambda e^{tu}.$$

For one thing, the convolution corresponds to the sum.

By the action b that is the translation, we have a change of intensity;

$$\psi(g(1,b)z) = e^{bu}\lambda e^{tu}.$$

The parameter b can be moved quite freely, differentiate with respect to the variable b and evaluate at $b = 0$. Repetitions make $u^p, p \geq 1$ and polynomials in u as well, so far as they are positive. They approximate positive continuous function. We are thus led to positive generalized functions on $(0, \infty)$. We wish to have the class of generalized functions as wide as possible.

Now we use the theorem on positive generalized functions (distributions in the Schwartz sense) to conclude that the space of positive generalized functions should be a measure (L. Schwartz). Thus we can state

Theorem 6.2. *The translation of the variable t makes compound Poisson process remain within the unit multiplicity.*

We have a freedom to choose λ in the construction of noises in Case II and III. It is reasonable to let λ be a space parameter that plays similar but with different roles. With this aim, we shall play the same game as in the case of the time parameter but we form independent system. This can be done in the following manner.

Take $[0,1]$ as in i) and ii), and form a product set $[0, 1] \times [0, 1]$ it is a set of (t, λ). With λ direction we associate independent random variables Y_n^k as X_n^k in the case ii). Of course X_n^k's and Y_n^k's are all independent. Then the limit will define 2-dimensional parameter Poisson noise where both directions are independent. There λ direction is with space parameter.

The above discussion, we actually see the class of the intensity "as large as possible". But, so far as continuous set of intensities is concerned single cyclic subspace (unit multiplicity) is best possible. Higher multiplicity may occur by the effect of the dilations as was seen before.

Note 1. It is reminded that we have so far discussed positive summarized sequences, normalization of which give probability distributions, respectively. Such a modification has no influence on multiplicity.

Note 2. Multiplication of the intensity has been obtained by the parametrization using the scale parameter u.

Next, we may repeat the operation to (2) by changing the amount of the shift and choosing different u's, as many times as we wish, we can conclude

Theorem 6.3. *The operations $m(g(0,b)t)$ coming from the shift of t generate a measure, denote it by $dn(\lambda)$ sitting in front of e^{tu} of the generating function.*

The factors e^{tu} with λ generate a measure which may be denoted by $dn(\lambda)$.

The measure is decomposed into two parts: continuous part dn_0 and discrete parts $n_d(\lambda_k)$, which is countable.

It is easily seen that every discrete point λ_k admits the multiplicity (> 1), produced by dilations. While each λ in the continuous part produces infinitesimal random variable of Poisson type, so that further consideration will be done in this direction.

It seems to be better to express the measure dn_0 as $dn_0(\lambda(u))$ or simply by $dn_0(u)$ because the translation of t by b always in the form bu which gives the factor e^{bu}. Such a notation helps us to understand the decomposition of a Lévy process. Namely, we finally have the following theorem.

Theorem 6.4. *The general form of the ψ-function that admits invariance under the Affine transform is expressed in the form*

$$\psi(z) = \int dn_0(u)(e^z u - 1) + \sum_k \lambda_k \left(\sum_j e^{zu_{j,k}} \right).$$

Here is noted that $dn(u)$ involves a discrete part that produces the multiplicity.

In fact we need much study for the facts related to Theorem 6.4. We hope that we can continue the problem related to Theorem 6.4.

References

1. O. Bratteli and D.W. Robinson, Operator algebras and quantum statistical mechanics. vol.1, Springer, 1987.

2. B.V. Gnedenko and A.N. Kolmogorov, Limit distributions for sums of independent random variables. Translated from the Russian. Addison-Wesley Pub. Co. 1954.

3. T. Hida, Canonical representations of Gaussian processes and their applications. Mem. Coll. Sci. Univ. Kyoto, A 33, (1960), 109-155.

4. T. Hida and Si Si, Lectures on white noise functionals. World Sci. Pub. Co. 2008.

5. T.Hida and Si Si, White noise. World Sci. Pub. Co. to appear.

6. P.Lévy, Sur les intégrales dont les eleménts sont des variables aléatoires indépendants. Ann. della R. Scoula Normale Superior di Pisa. Ser. II, III (1934) 337 - 366. Also three C.R. papers (1934).

7. P. Lévy, Propriétés asymptotiques des summes de variables aléatoires indépendantes ou enchainées. Journ. de math. XIV, 1935, 347-402,

8. P. Lévy, Théorie de l'addition des variables aléatoires. Gauthier-Villars, 1937, 2éme ed. 1954,

9. P. L'evy, Processus stochastiques et mouvement brownien. Gauthier-Villars. 1948, 2éme éd. etaugmentée, 1965.

10. Si Si, Introduction to Hida distributions. World Scientific Pub. Co. 2012.

11. Si Si, New noise depending on the space parameter and the concept of multiplicity, IDAQP, 2014, Singapore Workshop paper.

Quantum Bio-Informatics VI

© 2020 World Scientific Publishing Co. Pte. Ltd.

pp. 163–176

ON TREATMENT OF GAUSSIAN COMMUNICATION
PROCESS BY QUANTUM ENTROPIES

NOBORU WATANABE* and TAKUMI MAKIWARA

Department of Information Sciences,
Tokyo University of Science,
Noda City, Chiba 278-8510, Japan
*E-mail: * watanabe@is.noda.tus.ac.jp*

Dedicated to Professor Masanori Ohya

In order to discuss the efficiency of information transmission of the Gaussian communication processes consistently, we introduced an entropy type functional and a
mutual entropy type functional in reference [1]. In that study, we used the set of
Gaussian input measures with covariance operators of trace one.

In this paper, we modify the conditions and generalize the entropy functional and
the mutual entropy functional to study more general input Gaussian measures and
Gaussian channels.

1. Introduction

In information theory, the efficiency of the communication processes
is discussed based on the Shannon entropy and the mutual entropy
(information), which satisfy the Shannon's type fundamental inequalities
such as

$$0 \leq I(p; \Lambda^*) \leq \min \{S(p), S(\Lambda^* p)\}.$$

The above inequalities show that the mutual entropy denotes the amount of
information correctly transmitted from the input system to the output system through the channel. The mutual entropy (information) in Gaussian
communication process is called the Kullback - Leibler information (mutual
entropy) [7]. If we assume the entropy of the input Gaussian probability measure is given by the discrete probability distribution concerning the set of
all finite measurable partitions of the initial space, then it becomes infinite
for all input Gaussian probability measures. If we suppose the entropy of
the input Gaussian probability measure is taken by the differential entropy,
then it does not always satisfy the Shannon's type fundamental inequalities

by means of the Kullback - Leibler information. In [16], we defined the entropy functional and the mutual entropy functional under the condition of trace one based on the quantum communication theory. In order to study the efficiency of the information transmission for the Gaussian communication processes, the treatments of that by means of the weak trace preserving condition are discussed in [22,23,24,25].

In this paper, we modify the conditions and generalize the entropy functional and the mutual entropy functional for the Gaussian communication processes.

2. Gaussian Communication Process

Based on [3,4,19,16], we review the Gaussian communication process.

Let $\mathcal{H}_1$ and $\mathcal{H}_2$ be real separable Hilbert spaces of an input and output systems. We denote the sets of all bounded linear operators and positive self-adjoint trace class operators on $\mathcal{H}_k$ by $\mathbf{B}(\mathcal{H}_k)$ and $\mathbf{T}(\mathcal{H}_k)_+$ (i.e., $\mathbf{T}(\mathcal{H}_k)_+ \equiv \{\rho \in \mathbf{B}(\mathcal{H}_k); \rho \geq 0, \rho = \rho^*, tr\rho < \infty\}$), respectively $(k = 1, 2)$. μ is called a Gaussian measure in $\mathcal{H}_k$ if μ is a Borel measure such that for $x \in \mathcal{H}_k$, there exist real numbers m_x and σ_x (> 0) satisfying

$$\mu\left(\{y \in \mathcal{H}_k; <y, x> \leq a\}\right) = \int_{-\infty}^{a} \frac{1}{\sqrt{2\pi\sigma_x}} \exp\left\{\frac{-(t - m_x)^2}{2\sigma_x}\right\} dt.$$

The characteristic function $\hat{\mu}$ of μ is given by

$$\hat{\mu}(x) = \exp\left\{i \langle x, m_x \rangle - \frac{1}{2} \langle x, R_\mu x \rangle\right\}, \quad R_\mu \in \mathbf{T}(\mathcal{H}_k)_+.$$

Let $\mathcal{B}_k$ be the Borel σ-field of $\mathcal{H}_k$ $(k = 1, 2)$ and μ be a Borel probability measure on $\mathcal{B}_k$ satisfying

$$\int_{\mathcal{H}_k} \|x\|^2 d\mu(x) < \infty.$$

For given μ, the mean vector $m_\mu \in \mathcal{H}_k$ and the covariance operator R_μ of μ are defined by

$$\langle x_1, m_\mu \rangle = \int_{\mathcal{H}_k} \langle x_1, y \rangle \mu(dy)$$

$$\langle x_1, R_\mu x_2 \rangle = \int_{\mathcal{H}_k} \langle x_1, y - m_\mu \rangle \langle y - m_\mu, x_2 \rangle \mu(dy)$$

for any $x_1, x_2, y \in \mathcal{H}_k$ $(k = 1, 2)$. We describe a Gaussian measure μ with a mean vector m and a covariant operator R by $\mu = [m, R]$. Let $\mathbf{P}(\mathcal{H}_k)$ be the set of all Borel probability measures on $\mathcal{H}_k$.

According to [3,4,19,16], we briefly explain a mathematical treatment of Gaussian communication process.

Let $(\mathcal{H}_1, \mathcal{B}_1)$ and $(\mathcal{H}_2, \mathcal{B}_2)$ be an input and output spaces, respectively. $\mathbf{P}_G^{(k)}$ is the set of all Gaussian probability measures on $(\mathcal{H}_k, \mathcal{B}_k)$ $(k = 1, 2)$. Let $\mu_0 \in \mathbf{P}_G^{(2)}$ be a Gaussian probability measure representing an additive noise of the Gaussian channel. For $\mu_1 \in \mathbf{P}_G^{(1)}$, a mapping Γ^* from $\mathbf{P}_G^{(1)}$ to $\mathbf{P}_G^{(2)}$ is defined by the Gaussian channel $\lambda : \mathcal{H}_1 \times \mathcal{B}_2 \to [0, 1]$ such as

$$\Gamma^* (\mu_1)(Q) \equiv \int_{\mathcal{H}_1} \lambda(x, Q) d\mu_1(x)$$
$$\lambda(x, Q) \equiv \mu_0 (\{y \in \mathcal{H}_2; \; Ax + y \in Q\}), \; x \in \mathcal{H}_1, \; Q \in \mathcal{B}_2,$$

where A is a linear transformation from $\mathcal{H}_1$ to $\mathcal{H}_2$, λ satisfies (1) $\lambda(x, \bullet) \in \mathbf{P}_G^{(2)}$ for each fixed $x \in \mathcal{H}_1$ and (2) $\lambda(\bullet, Q)$ is a measurable function on $(\mathcal{H}_1, \mathcal{B}_1)$ for each fixed $Q \in \mathcal{B}_2$. The compound measure μ_{12} by means of the input measure μ_1 and the output measure μ_2 is derived by

$$\mu_{12}(Q_1 \times Q_2) = \int_{Q_1} \lambda(x, Q_2) d\mu_1(x)$$
$$\mu_{12}(\mathcal{H}_1 \times Q_2) = \mu_2(Q_2)$$
$$\mu_{12}(Q_1 \times \mathcal{H}_2) = \mu_1(Q_1)$$

for any $Q_1 \in \mathcal{B}_1$ and $Q_2 \in \mathcal{B}_2$. Then the mutual entropy with respect to the input Gaussian measure μ_1 and the Gaussian channel λ is given by the Kullback - Leibler information such as

$$I(\mu_1; \lambda) = S(\mu_{12} | \mu_1 \otimes \mu_2)$$
$$= \begin{cases} \int_{\mathcal{H}_1 \times \mathcal{H}_2} \frac{d\mu_{12}}{d\mu_1 \otimes \mu_2} \log \frac{d\mu_{12}}{d\mu_1 \otimes \mu_2} d\mu_1 \otimes \mu_2 & (\mu_{12} \ll \mu_1 \otimes \mu_2) \\ \infty & \text{else} \end{cases}$$

where $\frac{d\mu_{12}}{d\mu_1 \otimes \mu_2}$ is the Radon - Nikodym derivative of μ_{12} with respect to $\mu_1 \otimes \mu_2$.

2.1. *Motivation of New Treatment for the Gaussian Communication Process*

For the input Gaussian probability measure μ_1, the differential entropy $S(\mu_1)$ is defined by

$$S(\mu_1) = - \int_{\mathbf{R}^2} \frac{d\mu_1}{dm} \log \frac{d\mu_1}{dm} dm,$$

where m is the Lebesgue measure and the entropy of the discrete probability distribution associated with the set of all finite partitions of the initial space is given by

$$S\left(\mu_1\right) = \sup\left\{-\sum_{A_k\in\tilde{A}}\mu_1\left(A_k\right)\log\mu_1\left(A_k\right);\quad \tilde{A}\in\mathcal{P}(\mathcal{B}_2)\right\},$$

where $\mathcal{P}(\mathcal{B}_2)$ is the set of all finite partitions of $\mathcal{B}_2$.

In [16], the differential entropy $S\left(\mu_1\right)$ of the input Gaussian measure μ_1 and the Kullback - Leibler information $I\left(\mu_1;\lambda\right)$ with respect to Gaussian probability measure μ_1 and the Gaussian channel λ hold

$$S\left(\mu_1\right) < I\left(\mu_1;\lambda\right).$$

It means that those entropic measures do not satisfy always the Shannon's type inequalities for a simple model of the Gaussian communication process. The entropy $S\left(\mu_1\right)$ of the discrete probability distribution associated with the set of all finite partitions of the initial space is always infinite. By using this entropy, it might be impossible to measure the difference in the Gaussian probability measures. In order to solve these inconsistent argument, we introduce a new treatment in [16] for the Gaussian communication process based on the quantum information communication theory.

2.2. Quantum Channels

Let $\mathbf{B}(\mathcal{H}_k)$ and $\mathfrak{S}(\mathcal{H}_k)$ be the set of all bounded linear operators and the set of all density operators on the complex separable Hilbert spaces $\mathcal{H}_k$ $(k = 1, 2)$, respectively. $(\mathbf{B}(\mathcal{H}_1), \mathfrak{S}(\mathcal{H}_1))$ and $(\mathbf{B}(\mathcal{H}_2), \mathfrak{S}(\mathcal{H}_2))$ denote the input and output quantum systems, respectively. A quantum channel Λ^* is a mapping from $\mathfrak{S}(\mathcal{H}_1)$ to $\mathfrak{S}(\mathcal{H}_2)$.

(1) A linear quantum channel Λ^* is the quantum channel satisfying the affine property (i.e., for any $\rho_k \in \mathfrak{S}(\mathcal{H}_1)$ and any nonnegative number $\lambda_k \subset [0, 1]$ with $\sum_k \lambda_k = 1$, $\Lambda^*(\sum_k \lambda_k \rho_k) = \sum_k \lambda_k \Lambda^*\left(\rho_k\right)$ holds.

(2) A completely positive (CP) channel Λ^* is the linear quantum channel and its predual map Λ from $\mathbf{B}(\mathcal{H}_2)$ to $\mathbf{B}(\mathcal{H}_1)$ of Λ^* satisfies

$$\sum_{i,j=1}^{n} A_i^*\Lambda(\overline{A}_i^*\overline{A}_j)A_j \geq 0$$

for any $n \in \mathbf{N}$, any $\{\overline{A}_i\} \subset \mathbf{B}(\mathcal{H}_2)$ and any $\{A_i\} \subset \mathbf{B}(\mathcal{H}_1)$, where the predual map Λ of Λ^* is defined by

$$tr\Lambda^*(\rho)B = tr\rho\Lambda(B)$$

for any $\rho \in \mathfrak{S}(\mathcal{H}_1)$ and any $B \in \mathbf{B}(\mathcal{H}_2)$.

The efficiency of the information transmission of the quantum communication processes described by the CP channels is investigated in [9], [6], [13], [14], [10,17,15].

2.2.1. *Quantum Communication Channel*

We briefly review the noisy quantum channel and the generalized beam splitter as examples of the quantum communication channels.

In order to discuss the quantum communication process considering noise and loss, the quantum channel of the quantum communication process with noise and loss was discussed in [10,15] adding the complex separable Hilbert space $\mathcal{K}_1$ of noise system and the complex separable Hilbert space $\mathcal{K}_2$ of loss system to the input and output systems.

Noisy Quantum Channel and Generalized Beam Splitter Let V be a linear mapping from $\mathcal{H}_1 \otimes \mathcal{K}_1$ to $\mathcal{H}_2 \otimes \mathcal{K}_2$ defined by

$$V(|n_1\rangle \otimes |m_1\rangle) = \sum_j^{n_1+m_1} C_j^{n_1,m_1}|j\rangle \otimes |n_1 + m_1 - j\rangle$$

for the photon number state vectors $|n_1\rangle \in \mathcal{H}_1, |m_1\rangle \in \mathcal{K}_1, |j\rangle \in \mathcal{H}_2, |n_1 + m_1 - j\rangle \in \mathcal{K}_2$, and $C_j^{n_1,m_1}$ is given by

$$C_j^{n_1,m_1} \tag{1}$$
$$= \sum_{r-L}^K (-1)^{n_1+j-r} \frac{\sqrt{n_1!m_1!j!(n_1 + m_1 - j)!}}{r!(n_1 - j)!(j - r)!(m_1 - j + r)!}$$
$$\times \alpha^{m_1-j+2r} \left(-\bar{\beta}\right)^{n_1+j-2r}$$

where α and β are complex numbers holding $|\alpha|^2 + |\beta|^2 = 1$ and $K = \min\{n_1, j\}$, $L = \max\{m_1 - j, 0\}$. In [15], we introduced a generalized beam splitting Π^* by

$$\Pi^*(\rho \otimes \xi) \equiv V(\rho \otimes \xi)V^*,$$

for an input state ρ in $\mathfrak{S}(\mathcal{H}_1)$ and a noise state $\xi \in \mathfrak{S}(\mathcal{K}_1)$. By using Π^*, we introduced in [15] the noisy quantum channel Λ^* with a fixed noise state

$\xi \in \mathfrak{S}(\mathcal{K}_1)$ defined by

$$\Lambda^*(\rho) \equiv tr_{\mathcal{K}_2} \Pi^*(\rho \otimes \xi) = tr_{\mathcal{K}_2} V (\rho \otimes \xi) V^*. \tag{2}$$

The Steinspring - Sudershan - Kraus form of the noisy quantum channel Λ^* with the noise state $\xi = |m\rangle\langle m|$ is denoted by

$$\Lambda^*(\rho) = \sum_{i=0}^{\infty} O_i V Q^{(m)} \rho Q^{(m)*} V^* O_i^* \quad (\forall \rho \in \mathfrak{S}(\mathcal{H}_1))$$

where $Q^{(m)} \equiv \sum_{l=0}^{\infty} (|y_l\rangle \otimes |m\rangle)\langle y_l|$, $O_i \equiv \sum_{k=0}^{\infty} |z_k\rangle (\langle z_k| \otimes \langle i|)$, $\{|y_l\rangle\}$ is a CONS in $\mathcal{H}_1$, $\{|z_k\rangle\}$ is a CONS in $\mathcal{H}_2$ and $\{|i\rangle\}$ is the set of number states in $\mathcal{K}_2$. For the coherent input state $\sigma = |\theta\rangle\langle\theta| \otimes |\kappa\rangle\langle\kappa| \in \mathfrak{S}(\mathcal{H}_1 \otimes \mathcal{K}_1)$, the output state of Π^* for σ is obtained by

$$\Pi^*(|\theta\rangle\langle\theta| \otimes |\kappa\rangle\langle\kappa|) = |\alpha\theta + \beta\kappa\rangle\langle\alpha\theta + \beta\kappa|$$
$$\otimes \left|-\bar{\beta}\theta + \bar{\alpha}\kappa\right\rangle\left\langle-\bar{\beta}\theta + \bar{\alpha}\kappa\right|.$$

The generalized beam splitting Π^* with the vacuum noise state $\xi_0 = |0\rangle\langle 0|$ is called the beam splitter Π_0^* given by

$$\Pi_0^*(|\theta\rangle\langle\theta| \otimes |0\rangle\langle 0|) = |\alpha\theta\rangle\langle\alpha\theta| \otimes \left|-\bar{\beta}\theta\right\rangle\left\langle-\bar{\beta}\theta\right|$$

for the coherent input state $\rho \otimes \xi_0 = |\theta\rangle\langle\theta| \otimes |0\rangle\langle 0| \in \mathfrak{S}(\mathcal{H}_1 \otimes \mathcal{K}_1)$. The beam splitter Π_0^* was described by means of the lifting $\mathcal{E}_0^*$ from $\mathfrak{S}(\mathcal{H})$ to $\mathfrak{S}(\mathcal{H} \otimes \mathcal{K})$ in the sense of Accardi and Ohya [1] as follows

$$\mathcal{E}_0^*(|\theta\rangle\langle\theta|) = |\alpha\theta\rangle\langle\alpha\theta| \otimes |\beta\theta\rangle\langle\beta\theta|.$$

Based on the liftings, the beam splitting was studied by Accardi - Ohya and Fichtner - Freudenberg - Libsher [5]. Moreover, the noisy quantum channel Λ_0^* with the vacuum noise state $\xi_0 = |0\rangle\langle 0|$ is called the attenuation channel given by Ohya [10] as

$$\Lambda_0^*(\rho) \equiv tr_{\mathcal{K}_2} \Pi_0^*(\rho \otimes \xi_0) = tr_{\mathcal{K}_2} V_0(\rho \otimes |0\rangle\langle 0|) V_0^*, \tag{3}$$

which plays an important role for investigating the quantum communication processes.

3. Quantum Entropy

3.0.1. *von Neumann Entropy and Ohya mutual entropy*

We here review the von Neumann entropy and the Ohya mutual entropy.
The von Neumann entropy $S(\rho)$ [8] is defined by

$$S(\rho) = -tr\rho\log\rho$$

for any density operators $\rho \in \mathfrak{S}(\mathcal{H}_1)$, which satisfies (1) $S(\rho) \geq 0$, (2) if $\Lambda^* = id$ (id is the identity channel), then $S(\Lambda^* \rho) = S(\rho)$, (3) $S(\rho_1 \otimes \rho_2) = S(\rho_1) + S(\rho_2)$. The quantum mutual entropy with respect to the initial state ρ and the quantum channel Λ^* is introduced by Ohya [10] defined such as

$$I(\rho; \Lambda^*) \equiv \sup \left\{ \sum_n S(\sigma_E, \rho \otimes \Lambda^* \rho), \rho = \sum_n \lambda_n E_n \right\},$$

where σ_E is the compound state given by $\sigma_E = \sum_n \lambda_n E_n \otimes \Lambda^* E_n$ associated with the Schatten-von Neumann (one dimensional spectral) decomposition [18] $\rho = \sum_n \lambda_n E_n$ of the input state ρ, and $S(\cdot, \cdot)$ is the Umegaki's relative entropy [20] denoted by

$$S(\rho, \sigma) \equiv \begin{cases} tr\rho (\log \rho - \log \sigma) & (s(\rho) \leq s(\sigma)) \\ \infty & \text{otherwise} \end{cases} \quad (4)$$

where $s(\rho)$ is the support projection of ρ. It was extended to more general quantum systems by Araki and Uhlmann [2,11,12,13,21]. The Ohya mutual entropy $I(\rho, \Lambda^*)$ and the von Neumann entropy hold the Shannon's fundamental inequalities such as

$$0 \leq I(\rho, \Lambda^*) \leq \min \{S(\rho), S(\Lambda^* \rho)\}.$$

If $\Lambda^* = id$, then $I(\rho, id) = S(\rho)$ is satisfied.

4. A Treatment of Gaussian Communication Process by Using Quantum Entropies

In [16], we proposed a mathematical treatment of Gaussian communication process restricted to the Gaussian measure $\mu = [0, R]$ into the subset $\mathbf{P}_{G,1}^{(k)}$ with $tr R = 1$ of $\mathbf{P}_G^{(k)}$ and a transformation $\Gamma^* : \mathbf{P}_{G,1}^{(1)} \to \mathbf{P}_{G,1}^{(2)}$ associated with the Gaussian channel λ given by

$$(\Gamma^* (\mu_1))(Q) = \int_{\mathcal{H}_1} \lambda(x, Q) \mu_1(dx).$$

For any $\mu_1 = [0, \rho_1] \in \mathbf{P}_{G,1}^{(1)}$, the Gaussian measure $\Gamma^* (\mu_1)$ is obtained by

$$\Gamma^* (\mu_1) = [0, A\rho_1 A^* + R_0],$$

where R_0 is the covariant operator of $\mu_0 \in \mathbf{P}_G^{(2)}$ with $tr R_0 < 1$, and A satisfies $A^* A \leq I_1$. For any $B_k \in \mathbf{B}(\mathcal{H}_k)$ and any $\mu_k \in \mathbf{P}_{G,1}^{(k)}$ ($k = 1, 2$), there exists a bijection Ξ_k^* from $\mathbf{P}_{G,1}^{(k)}$ to $\mathfrak{S}\left(\mathcal{H}_k\right)$ given by

$$tr\Xi_k^*\left(\mu_k\right)B_k = \int_{\mathcal{H}_k}\langle\xi,\ B_k\xi\rangle\mu_k\left(d\xi\right).$$

We define $\Pi_\alpha^*:\mathfrak{S}\left(\mathcal{H}_1\right)\to\mathfrak{S}\left(\mathcal{H}_2\right)$ consisted of Ξ_1^*, Ξ_2^* and Γ^* by

$$\Pi_1^*\left(\rho_1\right) = \Xi_2^*\circ\Gamma^*\circ\left(\Xi_1^*\right)^{-1}\left(\rho_1\right)$$

for any $\rho_1\in\mathfrak{S}\left(\mathcal{H}_1\right)$.

$$\mu_1\in\mathbf{P}_{G,1}^{(1)}\left(\subset\mathbf{P}_G^{(1)}\right)\ \xrightarrow{\Gamma^*}\ \Gamma^*\left(\mu_1\right)\in\mathbf{P}_{G,1}^{(2)}\left(\subset\mathbf{P}_G^{(2)}\right)$$
$$\Xi_1^*\downarrow\qquad\qquad\qquad\qquad\downarrow\Xi_2^*$$
$$\rho_1\in\mathfrak{S}\left(\mathcal{H}_1\right)\left(\subset\mathbf{T}(\mathcal{H}_1)_+\right)\ \xrightarrow{\Pi_1^*}\ \Pi_1^*\left(\rho_1\right)\in\mathfrak{S}\left(\mathcal{H}_2\right)\left(\subset\mathbf{T}(\mathcal{H}_2)_+\right)$$

For any $\rho_1\in\mathfrak{S}\left(\mathcal{H}_1\right)$, $\Pi_1^*\left(\rho_1\right) = A\rho_1 A^* + R_0$ is a positive trace class operator in $\mathcal{H}_1$. In [16], we suppose that $A^*A = \left(1 - trR_0\right)I_1$. It means that Π_1^* is a completely positive channel from $\mathfrak{S}\left(\mathcal{H}_1\right)$ to $\mathfrak{S}\left(\mathcal{H}_2\right)$. In the previous discussion [16], we assume the following three conditions:

- **Treatment $\mathbf{I}_1$** $\left(\Gamma^*:\mathbf{P}_{G,1}^{(1)}\to\mathbf{P}_{G,1}^{(2)}\right)$
  (1) Linearity condition (linear approximation)
  (2) Trace preserving condition with 1
  (3) Normality condition (trace of covariance operator is equal to one).

The treatment of Gaussian communication by using the weak trace preserving condition is studied in [22,23,24,25].

- **Treatment II** $\left(\Gamma^*:\mathbf{P}_G^{(1)}\to\mathbf{P}_G^{(2)}\right)$
  (1) Linearity condition (linear approximation)
  (2) Weak Trace preserving condition : $tr\Pi^*\left(R\right) = tr\Pi^*\left(R'\right)$ is hold for any $R, R'\in\mathbf{T}(\mathcal{H}_1)_+$ satisfying $trR = trR'$

In this paper, we generalize this condition such as

$$A^*A = \left(\alpha - trR_0\right)\frac{I_1}{\alpha}\quad\left(\alpha > 0\right).$$

We consider a mapping $\Pi_\alpha^*:\mathbf{T}(\mathcal{H}_1)_{+,\alpha}\to\mathbf{T}(\mathcal{H}_2)_{+,\alpha}$ consisted of Ξ_1^*, Ξ_2^*, Γ^* defined by

$$\Pi_\alpha^*\left(R_1\right) = \Xi_2^*\circ\Gamma^*\circ\left(\Xi_1^*\right)^{-1}\left(R_1\right),\quad\left(\forall R_1\in\mathbf{T}(\mathcal{H}_1)_{+,\alpha}\right).$$

In this paper, we suppose two conditions.

- **Treatment $\mathbf{I}_\alpha$** $(\Gamma^* : \mathbf{P}_G^{(1)} \to \mathbf{P}_G^{(2)})$

  (1) Linearity condition (linear approximation)
  (2) Trace preserving condition with α

$$
\begin{array}{ccc}
\mu_1 \in \mathbf{P}_G^{(1)} & \xrightarrow{\Gamma^*} & \Gamma^*(\mu_1) \in \mathbf{P}_G^{(2)} \\
\Xi_1^* \downarrow & & \downarrow \Xi_2^* \\
R_1 \in \mathbf{T}(\mathcal{H}_1)_{+,\alpha} & \xrightarrow{\Pi_\alpha^*} & \Pi_\alpha^*(R_1) \in \mathbf{T}(\mathcal{H}_2)_{+,\alpha}
\end{array}
$$

For any $R_1 \ (= \Xi_1^*(\mu_1)) \in \mathbf{T}(\mathcal{H}_1)_{+,\alpha}$, $\Pi_\alpha^*(R_1) = AR_1A^* + R_0$ satisfies the trace preserving condition with $trR_1 = tr\Pi_\alpha^*(R_1) = \alpha$. We define a mapping Λ_α^* from $\mathfrak{S}(\mathcal{H}_1)$ to $\mathfrak{S}(\mathcal{H}_2)$ by

$$
\Lambda_\alpha^* = \frac{1}{\alpha} \circ \Pi_\alpha^* \circ \alpha
$$

associated with Π_α^*. Then Λ_α^* is a completely positive channel from $\mathfrak{S}(\mathcal{H}_1)$ to $\mathfrak{S}(\mathcal{H}_2)$ as follows:

Theorem 4.1. *For a positive number α $(0 < trR_0 < \alpha)$, if Λ_α^* is defined by*

$$
\Lambda_\alpha^* = \frac{1}{\alpha} \circ \Pi_\alpha^* \circ \alpha,
$$

then (1) Λ_α^* is linear and (2) it's predual map Λ_α satisfies (a) completely positivity and (b) $\Lambda_\alpha(I) = I$. Thus Λ_α^* is a completely positive channel from $\mathfrak{S}(\mathcal{H}_1)$ to $\mathfrak{S}(\mathcal{H}_2)$.

Proof. For any $\rho, \sigma \in \mathfrak{S}(\mathcal{H}_1)$ and any $\lambda \in [0,1]$, one has

$$
\begin{aligned}
\Lambda_\alpha^*(\lambda\rho + (1-\lambda)\sigma) &= \frac{1}{\alpha} \circ \Pi_\alpha^* \circ \alpha\,(\lambda\rho + (1-\lambda)\sigma) \\
&= \lambda A\rho A^* + (1-\lambda)A\sigma A^* + \frac{R_0}{\alpha} \\
&= \lambda\left(A\rho A^* + \frac{R_0}{\alpha}\right) + (1-\lambda)\left(A\sigma A^* + \frac{R_0}{\alpha}\right) \\
&= \lambda\Lambda_\alpha^*(\rho) + (1-\lambda)\Lambda_\alpha^*(\sigma)
\end{aligned}
$$

The predual map Λ_α from $\mathbf{B}(\mathcal{H}_2)$ to $\mathbf{B}(\mathcal{H}_1)$ of Λ_α^* is obtained as follows: For any $Q \in \mathbf{B}(\mathcal{H}_2)$ and any $\rho \in \mathbf{T}(\mathcal{H}_1)_{+,1}$,

$$tr\rho\Lambda_\alpha(Q) = tr\Lambda_\alpha^*(\rho) Q$$

$$= tr\frac{1}{\alpha}(A(\alpha\rho)A^* + R_0) Q$$

$$= trA\rho A^*Q + tr\frac{R_0}{\alpha}Q$$

$$= tr\rho\left(A^*QA + \frac{trR_0Q}{\alpha}I\right)$$

For any $Q \in \mathbf{B}(\mathcal{H}_2)$, Λ_α is denoted by

$$\Lambda_\alpha(Q) = \left(A^*QA + \frac{trR_0Q}{\alpha}I\right)$$

For any $n \in \mathbb{N}$, any $B_j \in \mathbf{B}(\mathcal{H}_1)$ and any $A_k \in \mathbf{B}(\mathcal{H}_2)$, one has

$$\sum_{k,j=1}^{n} B_k^* \Lambda_\alpha\left(A_k^* A_j\right) B_j$$

$$= \sum_{k,j=1}^{n} B_k^*\left(A^* A_k^* A_j A + \frac{trR_0 A_k^* A_j}{\alpha}I\right) B_j$$

$$= \sum_{k=1}^{n} A_k^* A B_k^* \sum_{j=1}^{n} A_j A B_j + \frac{1}{\alpha}\sum_{k,j=1}^{n} B_k^*\left(trR_0^{\frac{1}{2}} A_k^* A_j R_0^{\frac{1}{2}}\right) B_j$$

$$= \left(\sum_{k=1}^{n} A_k A B_k\right)^*\left(\sum_{j=1}^{n} A_j A B_j\right)$$

$$+ \frac{1}{\alpha}\sum_{p,q=1}^{n}\left(\sum_{k=1}^{n}\langle x_q| R_0^{\frac{1}{2}} |y_p\rangle B_k\right)^*\left(\sum_{j=1}^{n}\langle x_q| R_0^{\frac{1}{2}} |y_p\rangle B_j\right)$$

$$= C^*C + \frac{1}{\alpha}\sum_{p,q=1}^{n} D_{p,q}^* D_{p,q} \geq 0$$

Furthermore one has

$$\Lambda_\alpha(I) = \left(A^*A + \frac{trR_0}{\alpha}I\right)$$

$$= (\alpha - trR_0)\frac{I}{\alpha} + \frac{trR_0}{\alpha}I$$

$$= I$$

$\square$

For any $\mu_1 \in \mathbf{P}_G^{(1)}$, we denote the Schatten - von Neumann decomposition of R_1 by

$$R_1 = \sum_n \tau_n E_n,$$

where E_n is one dimensional orthogonal projection in $\mathfrak{S}(\mathcal{H}_1)$ and $\rho_1 = \sum_n \lambda_n E_n \in \mathfrak{S}(\mathcal{H}_1)$. We introduce the generalized Ohya compound state Σ_E with respect to $\Xi_1^*(\mu_1)$ and Λ_α^* and the trivial compound state Σ_0 given by

$$\Sigma_E = \sum_n \tau_n E_n \otimes \Lambda_\alpha^*(E_n)$$

$$= tr\Xi_1^*(\mu_1) \left[\sum_n \lambda_n E_n \otimes \Lambda_\alpha^*(E_n) \right]$$

$$= tr\Xi_1^*(\mu_1) \sigma_E \in \mathbf{T}(\mathcal{H}_1 \otimes \mathcal{H}_2)_+,$$

$$\Sigma_0 = \Xi_1^*(\mu_1) \otimes (tr\Xi_1^*(\mu_1)) \Lambda_\alpha^*(\rho_1)$$

$$= (tr\Xi_1^*(\mu_1))^2 \rho_1 \otimes \Lambda_\alpha^*(\rho_1) \in \mathbf{T}(\mathcal{H}_1 \otimes \mathcal{H}_2),$$

where σ_E is the Ohya compound state in respect to ρ_1 and Λ_α^*. We have the following lemma.

Lemma 4.1. *One has*

$$S(\Sigma_E, \Sigma_0)$$

$$= tr\Xi_1^*(\mu_1) S \left(\sum_n \lambda_n E_n \otimes \Lambda_\alpha^*(E_n), \rho_1 \otimes \Lambda_\alpha^*(\rho_1) \right)$$

$$-tr\Xi_1^*(\mu_1) \log tr\Xi_1^*(\mu_1).$$

Proof.

$$S(\Sigma_E, \Sigma_0) = S \left(tr\Xi_1^*(\mu_1) \sum_n \lambda_n E_n \otimes \Lambda_\alpha^*(E_n), (tr\Xi_1^*(\mu_1))^2 \rho_1 \otimes \Lambda_\alpha^*(\rho_1) \right)$$

$$= tr\Xi_1^*(\mu_1) S \left(\sum_n \lambda_n E_n \otimes \Lambda_\alpha^*(E_n), \rho_1 \otimes \Lambda_\alpha^*(\rho_1) \right)$$

$$-tr\Xi_1^*(\mu_1) \log \frac{(tr\Xi_1^*(\mu_1))^2}{tr\Xi_1^*(\mu_1)}$$

$$= tr\Xi_1^*(\mu_1) S \left(\sum_n \lambda_n E_n \otimes \Lambda_\alpha^*(E_n), \rho_1 \otimes \Lambda_\alpha^*(\rho_1) \right)$$

$$-tr\Xi_1^*(\mu_1) \log tr\Xi_1^*(\mu_1). \qquad \square$$

We introduce (1) the entropy type functional $\tilde{S}(\mu_1)$ of the input Gaussian measure $\mu_1 = [0, \Xi_1^*(\mu_1)]$ and (2) the mutual entropy type functional $\tilde{I}(\mu_1; \lambda)$ with respect to the input Gaussian measure μ_1 and the Gaussian channel λ such as

$$\tilde{S}(\mu_1) \equiv -tr\Xi_1^*(\mu_1)\log\Xi_1^*(\mu_1),$$
$$\tilde{I}(\mu_1; \lambda) \equiv \sup_E S(\Sigma_E, \Sigma_0).$$

Then one can obtain the following inequalities:

Theorem 4.2. *For any $\mu_1 \in \mathbf{P}_G^{(k)}$ and for some Gaussian channel λ, one has the Shannon's type fundamental inequalities as*

$$0 \le \tilde{I}(\mu_1; \lambda) \le \tilde{S}(\mu_1).$$

Proof. By using the above lemma and the Uhlmann's monotonicity, one has

$$S(\Sigma_E, \Sigma_0) = tr\Xi_1^*(\mu_1) S\left(\sum_n \lambda_n E_n \otimes \Lambda_\alpha^*(E_n), \rho_1 \otimes \Lambda_\alpha^*(\rho_1)\right)$$

$$- tr\Xi_1^*(\mu_1)\log tr\Xi_1^*(\mu_1)$$

$$= tr\Xi_1^*(\mu_1)\sum_n \lambda_n S(\Lambda_\alpha^*(E_n), \Lambda_\alpha^*(\rho_1)) - tr\Xi_1^*(\mu_1)\log tr\Xi_1^*(\mu_1)$$

$$\le tr\Xi_1^*(\mu_1)\sum_n \lambda_n S(E_n, \rho_1) - tr\Xi_1^*(\mu_1)\log tr\Xi_1^*(\mu_1)$$

$$= tr\Xi_1^*(\mu_1) S(\rho_1) - tr\Xi_1^*(\mu_1)\log tr\Xi_1^*(\mu_1)$$

$$= \tilde{S}(\mu_1)$$

Thus by taking the supremum for the Schatten decomposition of $\Xi_1^*(\mu_1)$, we have

$$\tilde{I}(\mu_1; \lambda) = \sup_E S(\sigma\Sigma_E, \Sigma_0) \le \tilde{S}(\mu_1). \qquad \square$$

References

1. L. Accardi and M. Ohya, Compound channels, transition expectation and liftings, Appl. Math, Optim., **39**, 33-59 (1999).
2. H. Araki, Relative entropy for states of von Neumann algebras, Publ. RIMS Kyoto Univ. **11**, 809–833, (1976).
3. C.R. Baker, Mutual information for Gaussian processes, SIAM J. Appl. Math., 19, 451-458, (1970).

4. C.R. Baker, Capacity of the Gaussian channel without feedback, Inform. And Control, 37, 70-89 (1978).

5. K.H. Fichtner, W. Freudenberg and V. Liebscher, Beam splittings and time evolutions of Boson systems, Fakultat fur Mathematik und Informatik, Math/ Inf/96/ **39**, Jena, 105 (1996).

6. R.S. Ingarden, A. Kossakowski and M. Ohya, *Information Dynamics and Open Systems*, Kluwer, (1997).

7. Kullback, S. and Leibler, R., On information and sufficiency, **22**, 79-86 (1951).

8. J. von Neumann, *Die Mathematischen Grundlagen der Quantenmechanik*, Springer-Berlin, (1932).

9. M. Ohya, Quantum ergodic channels in operator algebras , J. Math. Anal. Appl., **84**, 318-328, (1981).

10. M. Ohya, On compound state and mutual information in quantum information theory, IEEE Trans. Information Theory, **29**, 770-774 (1983).

11. M. Ohya, Note on quantum probability, L. Nuovo Cimento, **38**, 402-404, (1983).

12. M. Ohya, Some aspects of quantum information theory and their applications to irreversible processes, Rep. Math. Phys., **27**, 19-47 (1989).

13. M. Ohya and D. Petz, *Quantum Entropy and its Use*, Springer, Berlin, 1993.

14. M. Ohya and I. Volovich, *Mathematical Foundations of Quantum Information. and Computation and Its Applications to Nano- and Bio-systems*, Springer, (2011).

15. M. Ohya and N. Watanabe, Construction and analysis of a mathematical model in quantum communication processes, Electronics and Communications in Japan, Part 1, **68**, No.2, 29-34 (1985).

16. M. Ohya and N. Watanabe, A new treatment of communication processes with Gaussian channels, Japan Journal on Applied Mathematics, **3**, 197-206 (1986).

17. M. Ohya and N. Watanabe, *Foundation of Quantum Communication Theory (in Japanese)*, Makino Pub. Co., (1998).

18. R. Schatten, *Norm Ideals of Completely Continuous Operators*, Springer - Verlag, (1970).

19. A.V. Skorohod, *Integration in Hilbert space*, Springer, Verlag, Berlin New York, (1974).

20. H. Umegaki, Conditional expectations in an operator algebra IV (entropy and information), Kodai Math. Sem. Rep., **14**, 59-85 (1962).

21. A. Uhlmann, Relative entropy and the Wigner-Yanase-Dyson-Lieb concavity in interpolation theory, Commun. Math. Phys., **54**, 21–32, (1977).

22. N.Watanabe, On complexity of quantum communication processes, Americal Institute of Physics, 1508, 334-342, (2013).

23. N. Watanabe, An entropy based treatment of Gaussian communication process for general quantum systems, Open Systems and Information Dynamics, 20, No. 3, 1340009, 10 pp. (2013).

24. N. Watanabe, Entropy type complexity of quantum processes, Physica Scripta, 2014, No. T163 (2014).

25. N. Watanabe, Note on entropy-type complexity of communication processes, *WHITE NOISE ANALYSIS AND QUANTUM INFORMATION*, Edited by L.Accardi, Louis H.Y.Chen, T.Hida, M.Ohya, Si Si, N.Watanabe, Lecture Notes Series, Institute for Mathematical Sciences, National University of Singapore, World Scientific Publishing Co. Pte. Ltd. Vol.34, pp.215-pp.230, (2017).

Quantum Bio-Informatics VI
© 2020 World Scientific Publishing Co. Pte. Ltd.
pp. 177–184

ROTATION MECHANISM REVEALED FROM THE THREE DIMENSIONAL STRUCTURE AND SINGLE MOLECULE OBSERVATION OF V₁-ATPASE FROM *E. HIRAE*

ICHIRO YAMATO[1] and TAKESHI MURATA[2,3]

*[1]Department of Biological Science and Technology, Tokyo University of Science,
6-3-1 Niijuku, Katsushika-ku, Tokyo, 125-8585, Japan
[2]Department of Chemistry, Graduate School of Science, Chiba University,
1-33 Yayoi-cho, Inage, Chiba 263-8522, Japan
[3]JST, PRESTO, 1-33 Yayoi-cho, Inage, Chiba 263-8522, Japan*

Researches of biology are targeted on three major flows, materials (or chemicals), energy, and information. We have been mainly concerned with the studies on bioenergy transducing mechanisms. One of the authors (I. Yamato) has studied the mechanism of secondary active transport systems and proposed an affinity change mechanism as a general hypothesis, then tried to confirm that it is applicable to other kinds of bioenergy transducing systems. Choosing Na⁺-translocating V-type ATPase from *Enterococcus hirae* as a target, we hypothesized the affinity change mechanism for the energy transduction of this ATPase. Here we show several three dimensional structures of parts of the ATPase and single molecule observation studies supporting the hypothesis. From such detailed and extensive researches on protein structure/function relationship, we can proceed toward the *in silico* biology, which was described previously in 2007 ([1] "Toward *in silico* biology").

1. Introduction

One of the authors (I. Yamato) has summarized that life is created on the flows of materials, energy, and information [2]. We have been interested in the mechanism of energy flow in living systems. According to the extensive studies of secondary active transport mechanisms [3], Yamato has proposed a hypothesis that the energy input is mainly used for the affinity change of substrate [4]. Muscle is a typical energy transducing machinery in living systems, converting the chemical energy of ATP hydrolysis to exert forces. A thermal ratchet model for the energy transduction mechanism (chemical to mechanical energy) was proposed [5], where the energy input is used not for power stroke but for the choice of direction of movement and the driving force for the movement comes from the thermal fluctuation energy. But it is still difficult to obtain crucial evidence for or against the model. There is another kind of interesting machineries in living systems converting chemical energy (ATP hydrolysis) to mechanical energy (rotation); F-ATPases and V-ATPases. In the previous publication [2], Yamato introduced our research on the V-ATPase, a vacuolar-type ATPase translocating Na$^+$ inside of a bacterial cell, *Enterococcus hirae*, to outside (Fig. 1).

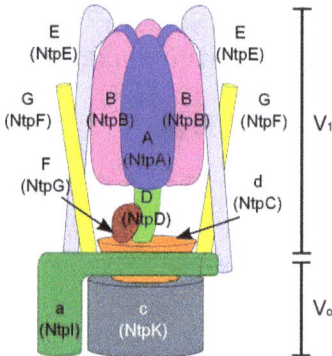

Fig. 1. A model of V-ATPase from *Enterococcus hirae*. V$_1$ indicates catalytic domain (consisting of A$_3$B$_3$DF), V$_o$ indicates membrane domain (consisting of a-c$_{10}$). Peripheral stalk consists of E and G subunits and central stalk consists of d, D, and F subunits (Parentheses show names of corresponding subunits of *E. hirae* V-ATPase).

Here we briefly introduce the *Enterococcus hirae* V-ATPase. *E. hirae* is a gram-positive fermentative bacterium and it can grow at high salt concentration and/or alkaline pH. The Na$^+$-translocating ATPase is responsible for this growth capability by pumping out Na$^+$ inside of cells [2, 6-8].

E. hirae V-ATPase is a homolog of eukaryotic V-ATPase which physiologically transports H$^+$ rather than Na$^+$ [7, 8]. This V-ATPase is composed of a soluble catalytic domain (V$_1$, A$_3$B$_3$DF) and an integral membrane domain (Vo, a-c$_{10}$) (Fig. 1) [9]. The central stalk of this ATPase is composed of d, D, and F subunits, where DF is called central axis. And the A$_3$B$_3$ part hydrolyses ATP to convert the chemical free energy to the rotation (torque) energy of the axis DF. The rotation of DF links to the rotation of c rotor ring in membrane embedded Vo part and Na$^+$ is transported through the interface of c and the stator, a. We have solved the 3D structures of our V-ATPase head part A$_3$B$_3$DF, the axis DF [10], and membrane embedded ion-translocating rotor ring c [11, 12]. Here we describe below the structure/function of the V$_1$ part converting ATP hydrolysis energy to the rotation of the axis.

2. Three dimensional structures of *E. hirae* V$_1$-ATPase and other subcomplexes

We have succeeded in crystallization of the V$_1$-ATPase (A$_3$B$_3$DF) and A$_3$B$_3$ complex. Figure 2 shows the three dimensional structures obtained [13]. They contained 2 nucleotide analogues (AMPPNP, adenilylimidodiphosphate), which are named bV$_1$ and bA$_3$B$_3$, respectively, or no nucleotide, which are named eV$_1$ and eA$_3$B$_3$.

Figure 2A shows the side view of eV$_1$ and 2B shows the top view. The top view shows 3 kinds of A$_1$B$_1$ motor units in the A$_3$B$_3$ motor part, empty-, bound- and tight-forms. In contrast to eV$_1$, eA$_3$B$_3$ (Fig. 2C; top view) has another set of 3 kinds of A$_1$B$_1$ units, empty-, bindable- and bound-forms. Even without nucleotide, eA$_3$B$_3$ has bound-form, showing an asymmetrical structure in spite of the assembly of the same A$_1$B$_1$ units. This asymmetrical structure explains well the unidirectional rotation caused by the order of ATP hydrolysis. And the binding of DF axis to eA$_3$B$_3$ induces the conformational change of eA$_3$B$_3$ to produce tight-form (Fig. 2B). V$_1$ even with bound nucleotides (bV$_1$) shows similar structure as eV$_1$, having empty-, bound- and tight-forms.

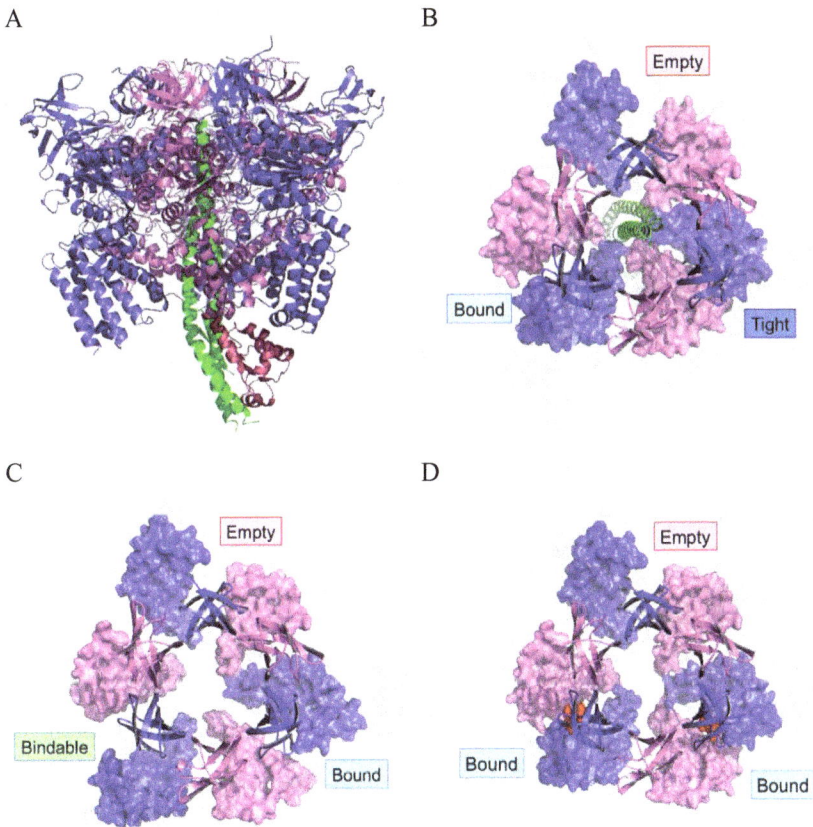

Fig. 2. Three dimensional structures of V$_1$ and A$_3$B$_3$ with and without bound nucleotides. A, side view of eV$_1$, drawn by ribbon representation. A subunit is drawn in blue, B subunit in magenta, D subunit in green and F subunit in orange. B, top view of eV$_1$, showing 3 kinds of A$_1$B$_1$ forms. N-terminal β-barrel part is drawn in ribbon representation, middle in transparent and C-terminal part in surface representation. C, top view of eA$_3$B$_3$. D, top view of bA$_3$B$_3$. Nucleotide (AMP-PNP) is drawn in red.

Figure 2D shows the top view of bA$_3$B$_3$, having further another set of 3 kinds of A$_1$B$_1$ units, empty-, bound- and bound-forms. Bindable-form in eA$_3$B$_3$ changes its conformation to bound-form induced by the nucleotide binding.

2.1. *ATP hydrolysis center at tight form*

From the structure, the catalytic center for ATP hydrolysis was predicted as at the tight-form in V$_1$. Without DF axis (bA$_3$B$_3$ having empty-, bound- and bound-forms), the binding site of AMPPNP between A and B subunits is rather open (Fig. 3), having the Arg-finger, which is known to be the responsible residue for the ATP hydrolysis activity, from B subunit at some distance from the γ-phosphate bond of the nucleotide. But the binding of DF axis induces a conformational change to tight-form in bV$_1$ having empty-, bound- and tight-forms. At the nucleotide binding site of the tight-form, the Arg-finger from B subunit comes closer to the γ-phosphate bond of the nucleotide, where the hydrolysis of ATP to produce ADP and phosphate is expected to take place.

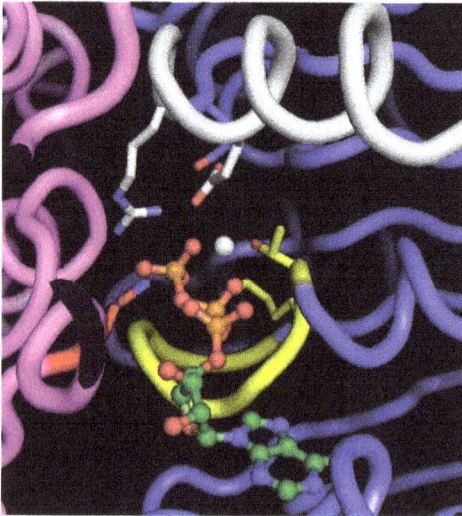

Fig. 3. The AMPPNP binding site between A and B subunits. A subunit is drawn in blue ribbon representation and B in magenta. The relevant amino acid residues for ATP hydrolysis are shown in stick model, with N atoms in blue and O atoms in red. P-loop, which is known to be the ATP binding site, in A subunit is drawn in yellow. The 'arm' region (fixed α-helix even at the conformational change; residues 261-275) is shown in white. AMPPNP is drawn in ball and stick representation, where C atoms in green and P atoms in orange. Arg-finger from B subunit shown in firebrick with blue N atoms locates near the γ-phosphate group in AMPPNP. The picture shows the structure of bound-form in bA$_3$B$_3$. Binding of DF axis to this form induces the conformational change to produce tight-form, where the Arg-finger tip comes closer to the γ-phosphate bond.

At present, there is no evidence for this speculation that the tight-form is the ATP hydrolysis center. Biochemical study is awaited for the confirmation of this model.

3. Demonstration of V₁-ATPase rotation by single molecule observation

The hydrolysis of ATP produces a torque for rotation of DF axis. The energy transduction mechanism of the ATP chemical energy converted to the mechanical force of rotation has not been fully elucidated. Structural information on the V_1 rotary motor should give an important clue to elucidate the mechanism. In addition, the single molecule observation of the rotation is also useful to gain such information about the rotation mechanism. Thus we tried to see the rotation of the DF axis in our V_1-ATPase [14, 15].

Figure 4 shows the result of such observation. Left panel shows the setup of the system; V_1 was attached on a Ni-NTA glass plate through A_3B_3 motor ring and a gold particle as a probe for observation was attached to DF axis through Streptavidin-biotin complex. The rotation of DF axis in the A_3B_3 ring can be seen as the rotation of the probe under microscope.

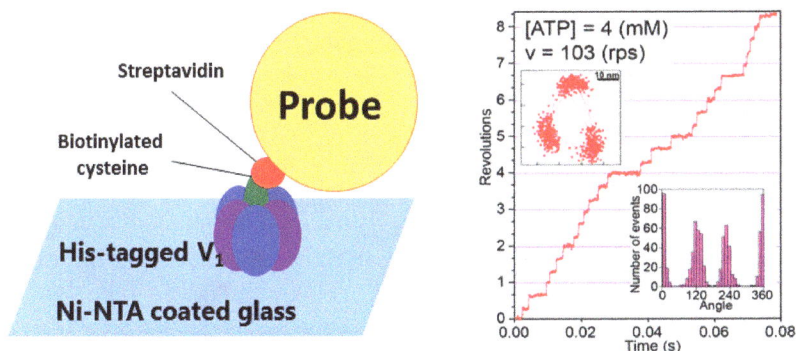

Fig. 4. Experimental setup for single molecule observation and the rotation of the V_1-ATPase [14]. Left, setup of the experiment. The A_3B_3 ring of the V_1-ATPase was fixed on the glass plate coated with Ni-NTA with N-terminal His$_6$ tag of the A subunits. Streptavidin-coated gold particle (probe) was attached to biotinylated cysteine residues in the rotor DF axis. Right, time evolution of the rotation. Rotation of the gold particle dependent on time is shown, speed of which was 103 rps as shown. Stepwise revolution was observable showing $120°$ rotation per step (3 steps for one cycle) as demonstrated in the left upper inset figure. The analysis of the angle distribution of the occupancy during rotation was shown in the right lower inset figure.

The result of the rotation was shown in Fig. 4 right panel. At 4 mM ATP, the axis rotated continuously with the rotation speed of 103 rps. The Km was about 100 μM. The angle distribution was analyzed showing the 3 step rotation for 1 cycle or $120°$ rotation per step. Different from the F_1-ATPase, the rotation did not show any sub-steps as $40°$ and $80°$ within the one step of $120°$. In F-ATPase, events of ATP binding, ATP hydrolysis, ADP release and phosphate release are thought to occur at $0°$, $200°$, $240°$ and $360°$. In our V-ATPase, it seems that ATP binding, ATP hydrolysis and ADP release occur at 0°, 240° and 240°, which may correspond to the empty-, tight- and tight-form. We have not yet examined the phosphate release step. The torque generation at continuous ATP hydrolysis state was estimated as about 20 pNnm [15], which is lower than those reported for F-ATPases.

For further detailed elucidation of the mechanism, further studies using biochemical approach, single-molecule observation, crystal structural approach and molecular dynamics simulation should be continued.

4. Rotation mechanism

At present, the mechanism of the rotation is not completely elucidated. But obtaining the three dimensional structures of A_3B_3 and V_1 crystals, we think we are now close toward the mechanism.

1. The structure obtained with the bound nucleotide (AMPPNP) of V_1 (similar as in Fig. 2B) can be designated as the state "waiting for ATP hydrolysis", as described in the above section 2. The ATP hydrolysis takes place at the tight-form in V_1, as described in the section 2.1. This event should corresponds to the 240° step as described in section 3. Produced phosphate is thought to be released easily, for which we have no data to prove.

2. Produced ADP should induce the conformational change of the whole V_1, which we speculate is the structure resembling the eA_3B_3 containing the bindable-form. The bindable-form should derive from the previous empty-form and the previous tight-form after ATP hydrolysis and ADP release gives rise the empty-form, finally making the structure resembling eA_3B_3. This structure should be the state "waiting for ATP binding".

3. The creation of the bindable-form derived from the empty-form caused by the ATP hydrolysis enables the binding of ATP to this site, making this bindable-form bound-form, producing the whole structure resembling the bA_3B_3 with bound-, bound- and empty-form; the last empty-form is speculated to derive from the tight-form after ATP hydrolysis and ADP release. The binding of ATP to the bindable-form corresponds to the event at 0° as described in section 3 and is speculated to induce a large conformational change to open the A_1B_1 unit to release the DF axis, that rotates clockwise by thermal fluctuation to the next bound-form to make it tight-form. Here, we further speculate that the rotation cannot be brought about by the force (torque) generation caused by the A_1B_1 conformational movement because if so, the DF release from the newly made empty-form and binding to the next A_1B_1 bound-form should occur by efficient exchange reactions of the binding interactions at the binding sites but it seems that the binding sites of neighboring two A_1B_1 units for DF axis are not close enough to be efficient for such seamless exchange, considering the three dimensional structures obtained. Therefore, it is conceivable that the release step to clockwise direction should necessitate a kind of energy (probably from the

ATP binding energy at the bindable-form) and the actual movement would be performed just by the thermal energy.

4. Then after release, rotation and binding of DF axis to the neighboring bound-form making it the tight-form, the whole structure resumes the initial bV_1 structure.

Acknowledgements

This work is supported by grants-in-aid for the Open Research Center Project (to I. Y.) and for Scientific Research (26291009), and Platform for Drug Discovery, Informatics, and Structural Life Science from MEXT.

References

1. I. Yamato, T. Ando, A. Suzuki, K. Harada, S. Itoh, S. Miyazaki, N. Kobayashi and M. Takeda (2008) Toward *in silico* biology (from sequences to systems). Quantum Bio-Informatics (From Quantum Information to Bio-Informatics) p.440-p.455 (eds L. Accardi, W. Freudenberg, & M. Ohya) (Proceedings of the International Symposium of Quantum Bio-Informatics Research Center 2007, Chiba) World Scientific, Singapore.
2. I. Yamato (2013) From structure and function of proteins toward *in silico* biology. Quantum Bio-Informatics V. (From Quantum Information to Bio-Informatics) p.473-p.485 (eds L. Accardi, W. Freudenberg, & M. Ohya) (Proceedings of the International Symposium of Quantum Bio-Informatics Research Center 2011, Chiba) World Scientific, Singapore.
3. I. Yamato (1992) Ordered binding model as a general mechanistic mechanism for secondary active transport systems. FEBS Lett., 298, 1-5.
4. I. Yamato (1993) Ordered binding model as a general tight coupling mechanism for bioenergy transduction --- A hypothesis. Proc. Japan Acad., 69, 218-223.
5. R. D. Vale and F. Oosawa (1990) Protein motors and Maxwell's demons: Does mechanochemical transduction involve a thermal ratchet? Adv. Biophys., 26, 97-134.
6. D. L. Heefner and F. M. Harold (1982) ATP-driven sodium pump in *Streptococcus faecalis*. Proc. Natl. Acad. Sci. USA, 79, 2798-2802.
7. Y. Kakinuma and K. Igarashi (1990) Release of component of *Streptococcus faecalis* Na$^+$-ATPase from the membranes. FEBS Lett., 271, 102-105.
8. Y. Kakinuma and K. Igarashi (1990) Some features of the *Streptococcus faecalis* Na$^+$-ATPase resemble those of vacuolar-type ATPase. FEBS Lett., 271, 97-101.

9. T. Murata, I. Yamato and Y. Kakinuma (2005) Structure and mechanism of vacuolar Na^+-translocating ATPase from *Enterococcus hirae*. J. Bioenerg. Biomembr., 37, 411–413.

10. S. Saijo, S. Arai, K. M. M. Hossain, I. Yamato, K. Suzuki, Y. Kakinuma, Y. Ishizuka-Katsura, N. Ohsawa, T. Terada, M. Shirouzu, S. Yokoyama, S. Iwata and T. Murata (2011) Crystal structure of the central axis DF complex of V-ATPase. Proc. Natl. Acad. Sci. USA, 108, 19955-19960.

11. T. Murata, I. Yamato, Y. Kakinuma, A. G. W. Lestlie and J. E. Walker (2005) Structure of the rotor of the V-type Na^+-ATPase from *Enterococcus hirae*. Science, 308, 654-659.

12. K. Mizutani, M. Yamamoto, K. Suzuki, I. Yamato, Y. Kakinuma, M. Shirouzu, J. E. Walker, S. Yokoyama, S. Iwata and T. Murata (2011) Structure of the rotor ring modified with N,N'-dicyclohexylcarbodiimide of the Na^+-transporting vacuolar ATPase. Proc. Natl. Acad. Sci. USA, 108, 13474-13479.

13. S. Arai, S. Saijo, K. Suzuki, K. Mizutani, Y. Kakinuma, Y. Ishizuka-Katsura, N. Ohsawa, T. Terada, M. Shirouzu, S. Yokoyama, S. Iwata, I. Yamato and T. Murata (2013) Rotation mechanism of *Enterococcus hirae* V_1-ATPase based on asymmetric crystal structures. Nature, 493, 703-707.

14. Y. Minagawa, H. Ueno, M. Hara, Y. Ishizuka-Katsura, N. Ohsawa, T. Terada, M. Shirouzu, S. Yokoyama, I. Yamato, E. Muneyuki, H. Noji, T. Murata and R. Iino (2013) Basic properties of rotary dynamics of *Enterococcus hirae* V_1-ATPase. J. Biol. Chem., 288, 32700-32707.

15. H. Ueno, Y. Minagawa, M. Hara, S. Rahman, I. Yamato, E. Muneyuki, H. Noji, T. Murata and Ryota Iino (2014) Torque generation of *Enterococcus hirae* V-ATPase. J. Biol. Chem., 289, 31212-31223.